ROUTLEDGE LIBRARY EDITIONS:
BUSINESS AND ECONOMICS IN ASIA

Volume 18

INNOVATION AND TECHNOLOGY TRANSFER IN JAPAN AND EUROPE

INNOVATION AND TECHNOLOGY TRANSFER IN JAPAN AND EUROPE

Industry-Academic Interactions

GLYN O. PHILLIPS

ASSISTED BY
MICHAEL HUGHES

Routledge
Taylor & Francis Group
LONDON AND NEW YORK

First published in 1989 by Routledge

This edition first published in 2019
by Routledge
2 Park Square, Milton Park, Abingdon, Oxon OX14 4RN

and by Routledge
52 Vanderbilt Avenue, New York, NY 10017

Routledge is an imprint of the Taylor & Francis Group, an informa business

British Library Cataloguing in Publication Data
A catalogue record for this book is available from the British Library

ISBN: 978-1-138-48274-6 (Set)
ISBN: 978-0-429-42825-8 (Set) (ebk)
ISBN: 978-1-138-36500-1 (Volume 18) (hbk)
ISBN: 978-0-429-43096-1 (Volume 18) (ebk)

Publisher's Note
The publisher has gone to great lengths to ensure the quality of this reprint but points out that some imperfections in the original copies may be apparent.

Disclaimer
The publisher has made every effort to trace copyright holders and would welcome correspondence from those they have been unable to trace.

Innovation and Technology Transfer in Japan and Europe

Industry-academic interactions

Glyn O. Phillips

assisted by
Michael Hughes

ROUTLEDGE
London and New York

First published 1989
by Routledge
11 New Fetter Lane, London EC4P 4EE
29 West 35th Street, New York NY 10001

Typeset by Pat and Anne Murphy, Highcliffe-on-Sea, Dorset
Printed in Great Britain by Billing & Sons Ltd, Worcester

British Library Cataloguing in Publication Data

Phillips, Glyn O. (Glyn Owain), *1927–*
Innovation and technology transfer in Japan and Europe: industry–academic interactions. 1. Great Britain. Industries. Technological innovation. 2. Japan. Industries. Technological innovation. I. Title. II. Hughes, Michael
338′.06

ISBN 0-415-03216-4

Library of Congress Cataloging in Publication Data

Phillips, Glyn O.
Innovation and technology transfer in Japan and Europe: industry–academic interactions/Glyn O. Phillips; assisted by Michael Hughes.
p. cm.
Bibliography: p.
Includes index.
ISBN 0-415-03216-4
1. Research, Industrial — Japan. 2. Research, Industrial — Great Britain. 3. Research, Industrial — European Economic Community countries. I. Title.
T177.J3P47 1989
338.95206 — dc20 89-31958 CIP

Dedicated to a true friend
Professor Yoshio Nakamura

Contents

Figures

Figures

Tables

Foreword

This book compares, contrasts, and evaluates the nature, scale, and direction of industry–academic interactions in Britain and Japan. Within this wider investigation, detailed consideration is given to the manner in which these interactions promote innovation and technology transfer.

Japan was chosen for study because its continuing success in world markets has been accompanied by the growing recognition that it is now a scientific, technological, and economic superpower. The advances Japan is making in R&D, technology transfer, and innovation are taking place at an impressive rate and set a standard against which our own efforts must be measured if European industry is to innovate, compete, and thrive against fierce international competition.

The book shows how Japan has developed particular arrangements for increasing and improving industrial–academic interactions to suit its own circumstances. There is no unique formula, and wholesale adoption of Japanese policies and systems is not the way forward. But as long as the Japanese economy enjoys the success that Britain and Europe need, we should be concerned with its mechanisms because they are so much a part of that success.

The information in this book provides a perspective against which we can take decisions about the future of our own industry–education interaction arrangements. There is much in this book for government, industrialists, and academics alike to study and evaluate. Specifically, the book should benefit the following: leaders of industry who undertake investment decisions, industrialists who wish to capitalize on academic advice and ideas, academics who wish to be informed about the various kinds of initiatives being taken, and central and local government agencies that wish to know how academic institutions can benefit local and regional economies.

To meet the above objectives the book is divided into four parts. The first part, comprising Chapters 1 and 2, assesses the background for industry–academic interaction. Chapter 1 compares R&D activity. Chapter 2 reviews the various organizations implementing R&D in Japan and Britain.

The second part, comprising Chapters 3 and 4, reviews current industry–academic arrangements. Chapter 3 reviews the systems and government policy towards interaction. Chapter 4 quantitatively assesses the degree of interaction in both countries.

The third part, comprising Chapters 5 and 6, presents case studies of industrial interaction with the academic community to detail some of the contemporary arrangements in Japan and Britain.

Part 4, comprising Chapter 7, examines European initiatives.

Finally, Chapter 8 evaluates and concludes by reference to the above chapters and two seminars organized by the authors in Japan to identify informed Japanese opinion.

The material collected in this volume has not previously been drawn together. The NEWTECH (Clwyd) Ltd. Business and Innovation Centre, in liaison with the Clwyd County Council, Welsh Development Agency (WINTECH) and the Nomura Research Institute, have supported the investigation in Japan and Britain. The author, however, is responsible for the views expressed.

Major emphasis is given to the work in Japan since such information is not readily available in Europe. Japanese collaboration was essential to collect the data since much of it is not even accessible in English. Our objective is to present an informed view of the complex yet successful Japanese system.

We hope it will benefit many like ourselves who have striven to keep abreast of the bewildering variety of programmes initiated in recent years in Europe and Japan to derive full economic benefit from the investment countries are making in higher education. We are particularly grateful to those Japanese industrial and academic specialists who freely participated in the dialogue and made the Japan seminars so successful. Their viewpoint cannot easily be obtained in Europe alone.

Glyn O. Phillips
Michael Hughes
The North East Wales Institute
Clwyd, Wales

Acknowledgements

I was privileged to visit Japan on a number of occasions during the preparation of this book, once for three months as Visiting Professor to the University of Gunma at Kiryu. I now have the opportunity to thank my many friends and scientific colleagues in Japan for their friendship, hospitality, and assistance. Nothing was too much trouble to be done on behalf of my wife, Rhiain, or myself. Our programme was meticulously arranged via Shinkansen and by personal stewardship to universities, industries, and central and local government research laboratories.

Rhiain recorded her impressions in the Welsh book *Yng Ngwyll Nihon* (Gwasg Gee). It explains that as westerners we see Japan only through a twilight, because of our inability to experience the Japanese culture directly through their language and literature. It will be apparent from the material recorded in this book how dependent I was on Japanese associates to obtain some awareness of their systems. Dr. T. Ohtsubo of Nomura Research Institute at Kamakura detailed the industrial–academic programmes and gave his time selflessly to assist my understanding of them. Dr. and Mrs. H. Hatakeyama (both distinguished scientists) introduced me to the wonders of Tsukuba Science City and their own home. Dr. Katsuyoshi Nishinari too arranged private visits and meetings of the Polymer Society and Society of Food and Agriculture on Industrial and Food Hydrocolloids, where I could meet a wide range of scientists. The Society of Fiber Science and Technology of Japan honoured me at their Hakone Symposium and by jointly organizing Cellucon-88 Japan. Their co-operation knew no bounds.

My main guide and mentor was Professor Yoshio Nakamura, Department of Chemistry, University of Gunma. Together we set out to build a bridge between Wales and Japan. His associated staff and students applied themselves with energy towards this end and our existing Mombusho-supported programme demonstrates that the bridge is now in place. The increasing two-way traffic may well require another very soon! Dr. T. Hongu of the Nisshinbo Company built a large section of the bridge himself. We thank the late Professor Ono (past president), and Professor T. Maekawa (present president of Gunma University), and Professor M. Suzuki, chairman of the Gunma University International Committee, for supporting all our efforts.

Clwyd County Council, through the chief executive, Mr. M. H. Phillips, has also supported this research and our link programmes. With help from the County Council, the Welsh Development Agency (WINTECH), Professor Nakamura, and Brother Industries, two seminars on Industrial–Education Co-operation were organized in Tokyo and Nagoya, enabling us to obtain the views of leading Japanese academics and industrialists. We include in Appendix II and III two outstanding contributions: from the industrial side Dr. Makato Kikuchi, Managing Director, Sony Research Center, Yokohama; from the academic standpoint, Professor Masamitsu Ohta, President, Nagoya Institute of Technology. No European could possibly provide this perspective on industry–academic associations in Japan. I sincerely thank them for their frank and authoritative papers.

The shortcomings and omissions can be attributed only to myself. To our many friends in Japan we say *Campai*, and please visit us in Wales soon.

Glyn O. Phillips

Organizations, acronyms and abbreviations

Japan

AIST	Agency of Industry, Science & Technology
FARTUN	Foundation for Applied Research at the Technological University of Nagaoka
FY	Fiscal Year
GIRIO	Government Industrial Research Institute, Osaka
JICA	Japan International Co-operation Agency
JITA	Japan Industrial Technology Association
JSPS	Japan Society for the Promotion of Science
MAFF	Ministry of Agriculture, Forestry & Fisheries
MCLSRL	Mitsubishi Chemical Life Science Research Laboratories
MHW	Ministry of Health & Welfare
MITI	Ministry of International Trade & Industry
MOE	Ministry of Education and Culture (Mombusho)
NASDA	National Space Development Agency
NCLI	National Chemical Laboratory for Industry
OMTRI	Osaka Municipal Technical Research Institute
RDCJ	Research Development Corporation of Japan
RSRD	Report on the Survey of R&D
SCRC	Sagami Chemical Research Center
STA	Science and Technology Agency
TRC	Tsukuba Research Consortium
TSF	Toray Science Foundation

Britain and Europe

ACARD	Advisory Council for Applied Research & Development
AFRC	Agricultural and Food Research Council
BIC	Business Innovation Centre

BRITE	Basic Research in Industrial Technologies
BTG	British Technology Group
CASE	Co-operative Awards in Science and Engineering
CDRA	Federation of Technology Centres
CERN	European Organization for Nuclear Research
CSP	Cambridge Science Park
DES	Department of Education and Science
DOEn	Department of Energy
DTI	Department of Trade and Industry
EBN	European Business and Innovation Centre Network
EC	European Community
ESPRIT	European Strategic Programme of R&D in Information Technology
EVCA	European Venture Capital Association
GDP	Gross Domestic Product
HEI	Higher Education Institutions
ICRO	Independent Contract Research Organizations
ICONE	Comparative Index of Norms in Europe
IKBS	Intelligent Knowledge Based Systems
ITeC	Information Technology Centres
MAFF	Ministry of Agriculture, Food and Fisheries
MMI	Man/Machine Interface
MOD	Ministry of Defence
MRC	Medical Research Council
NATO	North Atlantic Treaty Organization
NERC	Natural Environment Research Council
NEWI	North East Wales Institute of higher education
OECD	Organization for Economic Co-operation & Development
PERA	Production Engineering Research Association
SERC	Science and Engineering Research Council
SOEC	Statistical Office of the European Communities
STECLA	Standing Technological Conference of European Local Authorities
SUIC	Salford University Industrial Centre Ltd
TC	Technology Clubs
TCS	Teaching Company Scheme
TII	European Association for the Transfer of Technologies, Innovation and Industrial Information
UDIL	University Directors of Industrial Liaison
UGC	University Grants Committee
UKAEA	United Kingdom Atomic Energy Authority
VSLI	Very Large Scale Integration

Introduction

Successive governments have bemoaned the inability of British scientists to commercially exploit their research. On more than one occasion in Japan we were asked, 'How could a country that has won so many Nobel Prizes be so "poor"?' There was no criticism of the quality or the achievements of British research. Indeed, it was generally held in very high esteem. Might it therefore not be the fault of industry in failing to capitalize on the achievements of academics? The 1980s were well on their way before it was realized in Europe that interaction between industry and universities would not occur spontaneously. Mechanisms and incentives were needed to promote collaboration.

In Japan too, hardly anyone is satisfied with the degree or systems of co-operation between universities, their related institutes, and industry. This, however, should not be taken to mean that academics and their research do not contribute significantly to technological developments in Japan. That would not be true.

In recent years a bewildering variety of instruments and programmes have been generated in Europe and Japan to facilitate innovation and technology transfer within academic institutions and in association with industry. This book draws these initiatives together and provides a comparison of Japanese and British systems. Where European-wide programmes embrace Britain, these are included also. Indeed, it is the increasing industrial threat from Japan that has stimulated the European Commission to act.

Certain barriers have stood in the way of formal academic–industrial liaison in Japan, which did not even begin until 1965 or so. Student protests in the late 1960s and later were directed against professors receiving money from industry for their work. Throughout the western world at the time, it was the era of student protests; in Japan the emerging industrial giants were the targets. As a result, Mombusho, the rather bureaucratic and rigid Ministry of Education, which controls universities, has insisted on funding all research in the universities themselves. The Japanese professor, who is effectively a civil servant could not break this stranglehold. Until very recently, additional sources of finance were not welcome, even from other government agencies. The all-powerful Ministry of International Trade and Industry (MITI) was particularly regarded with suspicion. The effect of the isolation was the creation of a well-supported applied research machine in universities, with the basic requirements of people and equipment provided by Mombusho. The pinnacle of this system is the select group of ten or so national universities. The remainder of the 70 or so other universities, completely funded by government, have also sound research provision, but since they lack doctoral programmes do not attract the premier quality students. In a completely separate category are the private universities, which are a law unto themselves. They vary in quality and effectiveness.

Superimposed on the basic Mombusho research system is a new spirit of co-operation and entrepreneurial activity by the Japanese professor towards industry. Despite Mombusho, one way or other he has teamed up with industry, more often than not with the acquiescence of his own university authorities and backed by a local supportive industrial community. Even Mombusho has been forced to retreat before this pressure, with the result that a variety of systems for achieving co-operation have arisen, at national and prefectural levels.

Since the various procedures have not been set up systematically, it is not an easy task to detail them logically and simply. Yet their obvious benefits in technological development are so impressive that the effort to identify them must be worthwhile. The variety of government-directed and supported research initiatives, and the overall strength of the private research and development sector, jointly and individually influence universities. As a result the subject is indeed a complex one, as must be evident from the summary of the research system shown in the Figure below.

In industry, as in any athletic race, there are advantages in occupying second place. The front runner shields against prevailing winds; his or her tactical errors can readily be observed, and the decision when to kick for the tape rests solely with the challenger. For so long Japanese industry has thrived in the position of challenger; it is now in the unenviable position of being forced into the lead in many sectors, often against its own will.

Since they rejoined the industrial community in 1952, Japanese firms have spent more than $10 billion in fees and royalties for foreign technology. Now they are offering products for which there is little real competition in the west, particularly in consumer electronics.

SCIENTIFIC & TECHNOLOGICAL RESEARCH SYSTEM IN JAPAN

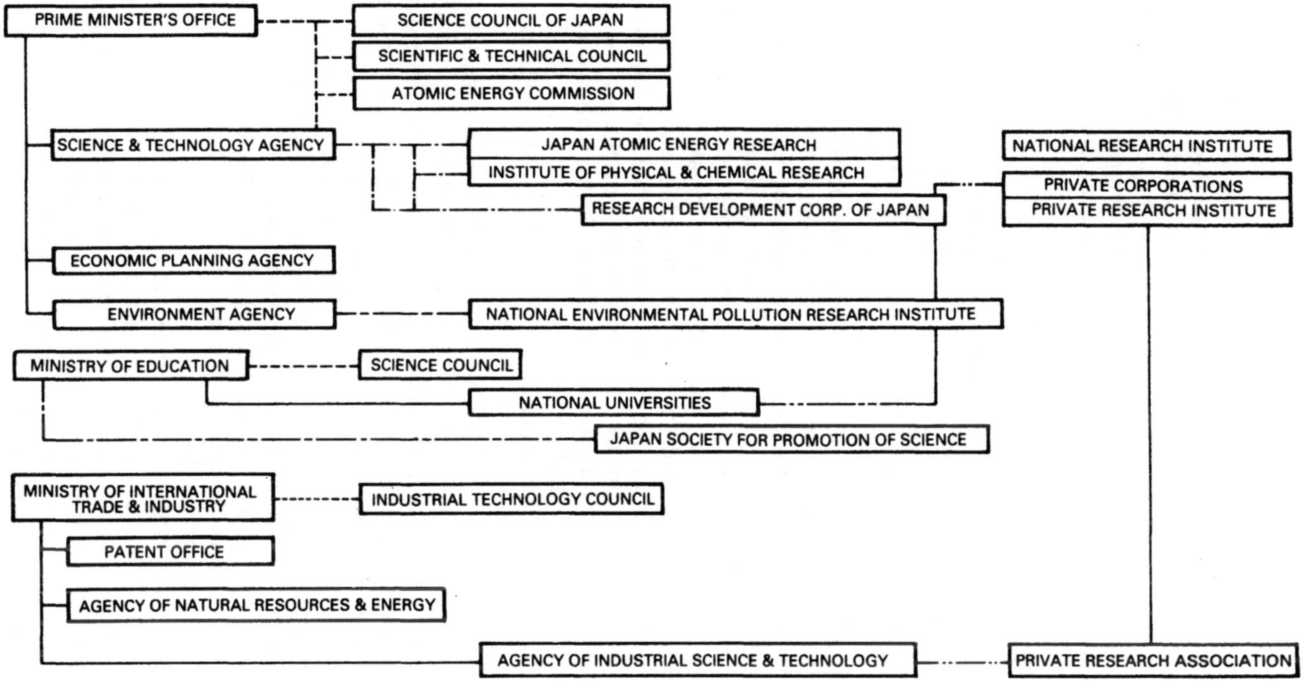

The explosive growth of this area provided a windfall for Japanese semiconductor manufacturers, and has given them a base that has now become a global presence. In polymers, pharmaceuticals, fermentation, genetic engineering and biotechnology, the growth has been equally startling for western competitors already staggering under the previous onslaught of the car, ship, and steel manufacturers.

These sectors now find themselves at the forefront of existing knowledge. Consequently, they can no longer count on the west to provide the scientific breakthroughs on which most of their current success is based. Indeed, the trend has already been reversed, for in 1978 Japan became a net exporter of new technology.

From now on the going will be very hard, and our visits to a wide range of Japanese industries confirmed that this is recognized by the forward planners. There is now a clear realization that copying and improving existing technologies is not enough.

Everyone believes that truly new discoveries are needed, but no one knows how they can be achieved. Indeed, at times we detected an attitude bordering on panic. Is there some secret means of achieving creativity? Can the truly new breakthroughs be managed? Will money and hard work alone achieve success?

Moreover, there is an underlying suspicion that the Japanese system and temperament are incompatible with original thinking. Undoubtedly this uncertainty about the future is percolating deeply into Japan's technological community.

Discoveries, of course, depend on individuals, not organizations, however well managed.

Could it be that in building its superb technological machine, Japan has neglected to nurture this creative activity? The race to pass examinations and cram information for school and university entrance might well exhaust the stamina of the young people for their more sensitive intellectual development. Conforming within a group may be incompatible with individual inspired thinking.

In any event, an overhaul of the existing ultra-rigid educational system can only benefit industry in the next century. Where are the ideas to be obtained for the period, say 10 years hence, when the now new developments such as the laser-based digital-audio disc players have become old hat?

The Ministry of International Trade and Industry's preference is obviously for a 'technology-based' nation described in its 'Vision of the 1980s'. Industry itself is rather less convinced that massive investment and concentration on narrow, individual targets such as high-performance ceramics or fifth-generation computers is the correct policy. It is doubtful whether even the collective mentality of the powerful MITI can pick a winner 10 years in advance.

Consequently, industry is turning to universities as a source of inspiration. These moves come after years of neglect of this relationship. Once this decision was made there was remarkable progress. The creation of the new Tsukuba Science City alone is a staggering affirmation of the faith now placed in industry–academic co-operation as the way forward to new ideas and products.

Economic survival was also the major motivation in Europe for pushing industry closer to the educational community. The year 1986 was designated Industry Year in Britain to make schools, colleges, and the community more aware of the role of industry in the community and to win acceptance for it. The explosive emergence of the Manpower Services Commission (now Training Agency) as a major force in schools, colleges, and among the adult unemployed has led to more emphasis on vocational education. Starving higher education institutions of core funding was undertaken with the direct objective of forcing them to work with industry, if only to secure the necessary money needed for their research survival. The Research Councils and government departments all detailed programmes and supplied the financial incentives to support the collaborations. Science, research, technology parks, and so on have been seen as the means of creating new high-technology 'sunrise' industries by harnessing the creative abilities of universities. To promote academic–industry interaction the resources of central and local government and the development agencies were increasingly directed towards joint programmes and the establishment of new innovative industries which involved and rewarded academics. A new 'enterprise culture' is being promoted throughout schools and colleges. This flood of 'enterprise' money has now been embraced by even the most traditional universities. The Enterprise in Higher Education programme initiated by the Training Agency in 1988 will provide £1 million over five years to 150 higher education institutions to integrate the concept of enterprise throughout the curriculum, and among all science, technology, and arts students.

Europe and Japan, therefore, albeit by different methods, are trying to commercialize the intellectual power of their young people. These fragmented and highly costly experiments are worthy of a detailed scrutiny, such as we have attempted in this book.

Part I

Research and development methods and expenditures

Chapter one

Research and development expenditure

International comparison

Table 1.1, using data compiled by the Organization for Economic Co-operation and Development (OECD), compares the R&D expenditure of Britain, Japan, and other countries which lead in R&D. The figures include, for all countries, government expenditure, both civil and defence, and industrially financed expenditure. The figures conceal important differences. For example, because of the arrangements following World War II, defence expenditure and hence R&D on defence is particularly low for Japan and West

Table 1.1 Gross and government-funded R&D expenditure on science, technology, social sciences, and humanities in 1983 for selected countries

	Gross R&D[b]		*Government-funded R&D*[c] *as a percentage of GDP*		
Country	*£ billion*	*As a percentage of GDP*	*Total*	*Defence*	*Civil*
Britain	6.7	2.3	1.34	0.66	0.68
France	7.1	2.1	1.40	0.46	0.94
Germany	9.9	2.6	1.15	0.11	1.04
Italy	3.2	1.2	0.71	0.04	0.67
Japan[a]	16.8	2.5	0.53	0.01	0.52
United States	48.7	2.7	1.18	0.76	0.42

Source: OECD, using OECD rates of exchange.

[a] 1982 for gross R&D, 1981 for government-funded R&D. There are considerable difficulties in comparing R&D funding by the Japanese government and funding by the government of other countries. For example, data available for Japan include only intramural spending and general university funding.

[b] Gross R&D expenditure, i.e. the sum of intramural R&D in all sections (government, higher education, private industry, and private non-profit).

[c] R&D expenditure, intramural and extramural, by government.

Germany compared with Britain, the United States, and France. Table 1.1 also shows government civil and defence R&D expenditure in relation to gross domestic product (GDP). There are, of course, difficulties associated with drawing international comparisons, and of ensuring comparability; the results must, therefore, be interpreted with caution. Nevertheless, the data indicate that British expenditure on government-funded R&D for defence is unduly high and that Japan spends twice as much on commercially productive R&D as Britain.

Trends

Japan's R&D has notably improved and expanded over the last 25 years, while R&D expenditure in Britain has fallen in real terms in recent years. Table 1.2 and Figure 1.1 show that Japanese R&D funds in fiscal year (FY) 1960 totalled ¥210.9 billion, representing only 1.15 per cent of the GDP. The corresponding figures for FY1983, however, were ¥7180.8 billion, equivalent to 2.58 per cent of the GDP.

Expenditure on R&D in Britain at current prices increased from £5.9 billion in 1981 to nearly £6.6 billion in 1983. However, after

Table 1.2 R&D trends in Japan

	Fiscal year 1960		*Fiscal year 1983*	
		NS[a]		*NS*[a]
R&D expenditure (¥ billion)	211	184	7,181	6,504
R&D expense/GNP	1.15	1.14	2.58	2.33
Elasticity of R&D expense on GNP[b]	1.22	1.21	2.38	2.14
Number of researchers	118	82	406	343
Industry	43	43	201	201
Research laboratory[c]	16	14	35	32
University	59	25	170	110
Number of research organizations	8,161	7,464	20,838	19,367
Industry	6,575	6,575	17,646	17,646
Research laboratory[c]	708	587	1,250	1,043
University	878	302	1,942	678

Source: Nomura Research Institute.

[a] *NS*, Natural sciences.

[b] $$\text{Elasticity of R\&D expense on GNP} = \frac{\text{Increasing rate of R\&D expense}}{\text{Increasing rate of GNP}}$$

[c] Research laboratory, including national laboratories, private laboratories and special corporations.

Figure 1.1 Growth of expenditure and persons employed on R&D in Japan

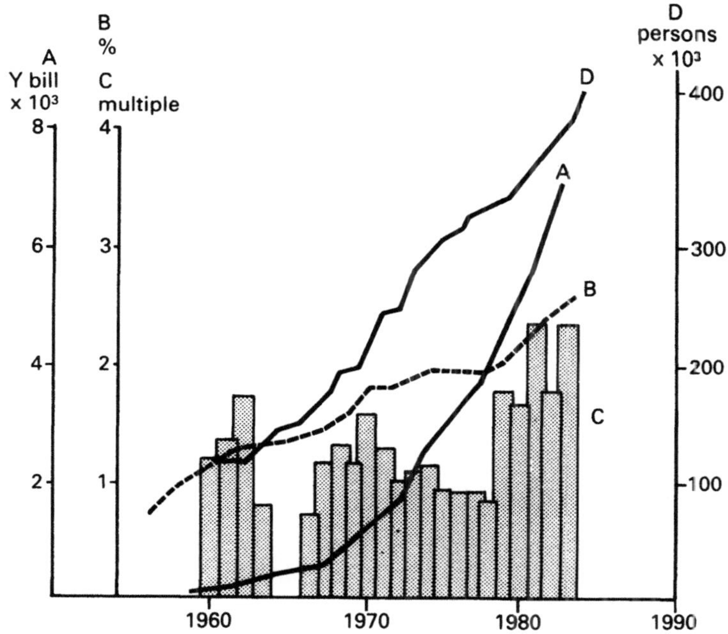

Source: Nomura Research Institute.

allowing for the increasing cost of R&D, expenditure fell in real terms by 4½ per cent. Over the same period GDP increased by more than 5½ per cent in real terms. Hence expenditure on R&D as a proportion of GDP fell from 2.34 per cent in 1981 to 2.19 per cent in 1983 (Table 1.3).

As the ratio of R&D investment to the GDP ranges from 2.0 to 2.7 per cent in the various western countries, Japan's R&D is certainly now conducted at a level of investment on a par with that of the United States, France, and Germany.

Japan's Agency of Industry, Science and Technology (AIST) has recommended that aggregate domestic R&D expenditure should be raised to 3 per cent of GDP. A picture is emerging, therefore, of Japan, with an already high R&D expenditure against GDP, only a negligible amount of which is spent on defence, proposing even further R&D expenditure.

Japan's investment in natural science R&D (physical science, engineering, agriculture, and medicine) totalled ¥6,503.7 billion in FY1983, an amount equivalent to 90.6 per cent of the ¥7,180.8

Table 1.3 Expenditure and employment on R&D in science and technology in Britain

	1972	*1975*	*1978*	*1981*	*1983*
Expenditure at current prices (£ millions)					
Analysis by performer					
By government	337	566	758	1,339	1,494
By universities[a] and further education establishments	115	179	317	630	750
By industry	831	1,340	2,324	3,792	4,163
By other sectors	30	66	111	160	176
Total	1,313	2,151	3,510	5,921	6,583
Analysis by source of funds[b]					
By government	640	1,117	1,651	2,826	3,218
By industry	571	877	1,552	2,529	2,869
By other sectors[c]	102	157	307	566	496
Total	1,313	2,151	3,510	5,921	6,583
Expenditure in 1975 prices (£ millions)[d]					
Government performed	575	566	511	596	574
Industry performed	1,418	1,340	1,566	1,661	1,564
Total[e]	2,242	2,151	2,365	2,603	2,486
Total expenditure as a percentage of GDP at market prices	2.07	2.06	2.14	2.34	2.19
Employment on R&D (end-year) (thousands)					
Central government	75.6	77.4	71.3	63.2	58.7
Industry	183.4	180.6	190.1	195.2	186

Source: HMSO, *Economic Trends*, C.S.O. no. 382 (August 1985).

Figures exclude research in the social sciences and humanities; excluding work performed abroad, figures relate to calendar years or corresponding financial years.

[a] Only very tentative estimates are available for this sector. Change in coverage between 1978 and 1981 tends to over-emphasize the apparent increase between these two years.

[b] As returned by performing sectors.

[c] Including overseas sector.

[d] Including R&D performed by other sectors.

[e] Current price data revalued using the SOEC (Statistical Office of the European Communities) price index of government performed R&D, and the Department of Trade and Industry ((DTI) R&D price index elsewhere.

billion total. This percentage of natural science R&D investment stood at 87.5 per cent in FY1960 and 89.3 per cent in FY1980, and has thus been steadily rising.

In the five-year period 1978 to 1983, Japan's total R&D funds increased at an annual average rate of 12.2 per cent in nominal terms and 7.2 per cent in real terms. The corresponding rates for natural science R&D funds were 12.8 and 7.8 per cent. The corresponding growth rates of Japan's GDP in nominal and real terms in the same period were 6.2 and 4.2 per cent respectively, indicating that R&D investment in Japan may be said to outstrip the growth of the GDP. Since this trend did not change even over the long term, the ratio of R&D investment to the GDP, as related above, climbed from 1.15 per cent in FY1960 to 2.58 per cent in FY1983.

Employment

In 1960 Japan had only 118,000 professional persons employed in research. By 1983 there were 406,000, representing a 3.4-fold increase. This figure rose further to 435,000 in 1984, a 3.7-fold increase over the 1960 total.

The total number of research personnel (including research assistants, technicians, and others) increased 2.3-fold from 317,000 in 1960 to 741,000 in 1983. The rate of increase in the number of research-related personnel is thus lower than that for professional researchers, indicating a qualitative improvement of research personnel viewed as a whole.

In Britain employment on R&D in science and technology, in terms of full-time equivalent worker-years, is shown at the bottom of Table 1.3. Employment on R&D by industry, after rising steadily to 1981, fell to below the 1978 level in 1983.

Establishments

The number of establishments implementing R&D in Japan grew from 9,700 in 1960 to 18,600 in 1983 and 20,800 in 1984, a 1.9-fold increase. The number of researchers per establishment is a rising trend. All figures quoted relate to the fiscal year.

Private investment on R&D

Some 76 per cent of Japan's R&D investment derives from funds in the private sector.

In Britain private industry provided about 35 per cent of the £6.6 billion spent on R&D in 1983 (£2,354 million) but achieved

Table 1.4 R&D performed in each sector according to the source of finance, Britain, 1983 (£ millions)

Sector providing the funds	*Government*	*Universities and further education establishments*	*Public corporations and research associations*	*Private industry*	*Other*	*Total performed in Britain*[a]	*Research carried out abroad*
Government	1,228.8	634.0	39.9	1,217.6	97.5	3,217.9	214.6
Universities[b]	—	25.0	—	—	—	25.0	—
Public corporations and research associations	12.6	9.1	433.5	60.0	—	515.2	—
Private industry	157.5	16.9	29.9	2,099.2	50.7	2,354.2	—
Overseas	42.1	18.0	23.0	260.3	4.5	347.9	—
Other	52.7	46.5	—	—	23.7	122.8	—
Total cost of research and development performed	1,493.8	749.5	526.2	3,637.1	176.4	6,583.0	—

Source: Department of Trade and Industry.

[a] Research in the social sciences is excluded.

[b] Including also further education establishments.

55 per cent (£3,637.1 million) of the work; most of the remainder was provided by the government. Of private industry's £2,354 million, £2,099 million was spent in its own establishments, £157 million in government establishments, £16.9 million in universities, and the rest in other establishments (Table 1.4).

It must not be overlooked that in Japan the ratio of R&D investment in the private sector to GDP reaches the highest level, even in the context of western countries. More specifically, Japan's private investment totalled 1.96 per cent of GDP, while the corresponding percentages for Britain, the United States, West Germany, and France were 1.3, 1.4, 1.6, and 0.9 per cent respectively.

Government expenditure on R&D

Japan

The Japanese government R&D budget is given in Table 1.5. The bulk of the R&D funds allocated to the Ministry of Education (MOE) are directed to universities (¥700 billion).

Table 1.5 Japanese government R&D budget — 1984 fiscal year (billions of yen)

	Government sources of R&D funds								
	Central government								
	MOE	*STA*	*MITI*	*MAFF*	*MOW*	*Others*	*Total*	*LG*	*Total*
Administration	3	1	3	1	—	2	11		11
GRI	8	31	41	[54]	[23]	[44]	[202]		202
SC	3	[334]	+	+	2	11	[351]		351
LGRI	—	—	—	3	—	—	3	[240]	240
International corporation	4	+	25	+	+	6	32		32
Universities	[700]	+	+	+	5	—	[704]		704
Commissioned research	+	+	[99]	1	+	+	102		102
Not specified	0	35	4	0	1	34	73		73
Totals	715	402	172	60	32	98	1,478	240	1,718

Sources: MOE, STA, MITI-AIST, MAFF, and MOW.

Notes: +, negligible plus; —, nil; AST, Agency for Science and Technology; GRI, National Laboratory; LG, local governments; LGRI, public laboratory; MAFF, Ministry of Agriculture, Fisheries and Forestry; MHW, Ministry of Health and Welfare; MITI, Ministry of International Trade and Industry; MOE, Ministry of Education; SC, special corporation; Others, other ministries not listed above.

R&D expenditure

The R&D funds for the Science and Technology Agency (STA) are directed primarily to the six special corporations under its control. These funds allocated to the special corporations are applied to work that must be conducted on a national scale, such as nuclear power research, space development, marine development, science and technology information centres, and the development of new technologies.

The budgets of the laboratories belonging to the various national ministries and agencies and those attached to local governments totalled ¥442 billion when taken together. The Ministry of Agriculture, Forestry, and Fisheries (MAFF), the Ministry of Health and Welfare (MHW), and other arms of the central government apply virtually all their R&D funds to the laboratories attached to them.

The Ministry of International Trade and Industry (MITI), however, uses its R&D funds in a variety of ways. In the first place, it allocated ¥41 billion for the 16 national laboratories attached to it, ¥25 billion for international research co-operation, and another ¥99 billion as funds for industrial technology policy having a national status. The main constituents of this industrial technology policy are (1) energy technology policy (the development of new energy technology, named the 'Sunshine Project', and the development of energy conservation technology, the 'Moonlight Project'), (2) policy for large-scale development of industrial technology, and (3) policy for next-generation industrial technologies. To meet its objectives, the Agency of Industrial Science and Technology (AIST) is given by MITI the functions of project organizer: allocating research to private corporations and forming project teams made up of members from industry, academia, and the government. The budget for this activity was ¥99 billion, as noted above.

The survey of the Japanese government R&D budget, Table 1.5 shows that the Ministry of Education (MOE), Space and Technology Agency (STA), MITI (AIST), and local governments play influential roles in the industry–academic interaction in Japan.

Britain

Tables 1.6 to 1.9 show total government expenditure on R&D. Table 1.6 analyses gross expenditure on R&D according to the programmes of the Public Expenditure White Paper (Cmnd 9428, I and II, HMSO). These are divided into R&D expenditure in the fields of science and technology, and the social sciences and humanities. Expenditure on science and technology was 95 per cent

Table 1.6 Gross central government expenditure on R&D in the financial year 1983, Britain (£ millions)

	Intramural expenditure					Extramural expenditure					
	Current										
	Salaries and wages[a]	*Other*	*Total*	*Capital*	*Total intramural*	*Universities and further education establishments*	*Private industry and public corporations*	*Other*	*Overseas*	*Total extramural*	*Total intra- and extramural*
R&D in science and technology											
Defence	366.1	200.7	566.8	88.1	654.9	9.4	1,257.3	7.0	97.9	1,371.7	2,026.5
Civil:											
Overseas services	3.1	1.8	4.9	0.2	5.1	2.7	0.1	3.1	7.5	13.4	18.5
Agriculture, fisheries and forestry	31.8	29.5	61.3	6.1	67.4	1.7	4.2	1.2	0.0[b]	7.1	74.5
Industry, trade, and employment											
Aerospace	0.5	0.1	0.6	—	0.6	0.7	145.4	—	51.1	197.2	197.8
Energy	169.6	160.4	330.0	35.2	365.1	6.3	32.1	7.1	2.5	48.0	413.1
Other industry, trade, and employment	18.0	14.1	32.1	5.0	37.1	2.3	111.3	5.5	0.3	119.4	156.5
Environmental services	16.9	9.2	26.1	2.0	28.1	3.8	20.2	3.7	0.2	27.9	56.0
Law, order, and protective services	2.7	3.0	5.7	0.1	5.8	0.3	2.6	0.1	0.2	3.2	9.1
Education and science											
Agricultural and Food Research Council	26.6	11.9	38.5	5.5	44.0	4.8	—	52.1	—	56.9	100.8
Medical Research Council	46.3	20.6	67.0	6.8	73.8	33.6	—	9.6	2.2	45.3	119.1
Natural Environment Research Council	33.3	26.2	59.5	9.7	69.2	7.1	0.4	9.8	0.0	17.3	86.5
Science and Engineering Research Council	35.0	35.3	70.3	21.8	92.1	78.8	0.1	30.2	52.6	161.7	253.8
Other science, education, etc.[c]	7.8	1.4	9.2	—	9.2	1.0	0.1	5.1	—	6.2	15.4
Universities	—	—	—	—	—	459.0	—	—	—	459.0	459.0
Health and social security	6.2	3.4	9.6	5.0	14.7	5.4	1.5	1.5	—	8.5	23.2
Other public and common services	0.4	0.7	1.1	0.1	1.1	0.0	0.1	—	—	0.1	1.2
Scotland	6.4	3.8	10.2	0.5	10.7	2.9	0.2	31.6	—	34.7	45.4
Wales	0.0	—	0.0	—	0.0	0.3	—	0.6	—	0.9	0.9
Northern Ireland	4.8	2.4	7.2	0.6	7.9	0.4	4.5	1.1	—	6.0	13.9
Total civil	409.3	323.9	733.2	98.7	831.9	611.1	322.8	162.4	116.6	1,212.9	2,044.8
Total R&D in science and technology	775.4	524.6	1,300.0	186.8	1,486.8	620.5	1,580.1	169.4	214.5	2,584.6	4,071.3

continued overleaf

Table 1.6 — continued

	Intramural expenditure					*Extramural expenditure*					
	Current										
	Salaries and wages[a]	*Other*	*Total*	*Capital*	*Total intramural*	*Universities and further education establishments*	*Private industry and public corporations*	*Other*	*Overseas*	*Total extramural*	*Total intra- and extramural*
Social science R&D[d]											
Defence	1.7	0.1	1.8	—	1.8	—	—	—	—	—	1.8
Civil:											
Overseas services	0.0	0.0	0.1	—	0.1	1.5	—	0.0	0.5	2.0	2.1
Agriculture, fisheries and forestry	—	—	—	—	—	0.2	—	0.0	—	0.2	0.2
Industry, trade, and employment	1.2	0.3	1.4	—	1.4	1.8	0.5	3.1	0.0	5.4	6.9
Environmental services	0.6	0.4	1.0	0.0	1.0	0.5	0.8	0.1	—	1.5	2.5
Law, order, and protective services	1.1	0.6	1.7	—	1.7	0.4	0.2	0.5	0.0	1.2	2.9
Education and science											
Economic and Social Research Council	2.0	1.2	3.2	0.0	3.3	9.6	—	4.4	—	14.0	17.3
Other science, education, etc.[a]	0.1	0.1	0.2	0.0	0.2	1.3	0.0	10.0	0.1	11.3	11.5
Universities	—	—	—	—	—	157.0	—	—	—	157.0	157.0
Health and social security	0.2	0.1	0.3	—	0.3	3.8	—	2.4	—	6.2	6.5
Other public and common services	1.4	1.0	2.4	—	2.4	0.2	0.1	0.0	0.0	0.4	2.8
Scotland	0.9	—	0.9	—	0.9	0.4	0.1	0.0	—	0.4	1.3
Wales	0.1	—	0.1	—	0.1	0.3	0.0	0.2	—	0.5	0.6
Northern Ireland	0.5	0.2	0.7	—	0.7	0.1	—	0.0	—	0.1	0.8
Total civil	8.2	3.8	12.0	0.1	12.1	177.0	1.8	20.9	0.6	200.3	212.3
Total social science R&D	9.9	3.9	13.8	0.1	13.8	177.0	1.8	20.9	0.6	200.3	214.1
Total expenditure on R&D	785.3	528.5	1,313.8	186.8	1,500.6	797.5	1,581.9	190.3	215.2	2,784.8	4,285.4

Source: Department of Trade and Industry.

[a] Total salary cost, including employer's national insurance contributions and superannuation payments.

[b] This notation is used throughout these tables to show numbers which, while not actually zero, are too small to register with the decimal places used.

[c] Including the Arts and Libraries programme.

[d] The figures for social science R&D are less reliable than those for R&D in general, because of the difficulties of distinguishing between social science research and other related activities such as information services and general purpose data collection.

Table 1.7 Current (non-capital) gross central government expenditure on intramural R&D and development, Britain (£ millions)

	1981–82[a]					1982–83[a]					1983–84				
	Basic research	*Applied research*	*Develop-ment*	*Un-allocated*	*Total*	*Basic research*	*Applied research*	*Develop-ment*	*Un-allocated*	*Total*	*Basic research*	*Applied research*	*Develop-ment*	*Un-allocated*	*Total*
Defence	—	176.1	329.3	—	505.3	—	209.2	344.9	—	554.1	—	232.8	344.0	—	566.8
Civil															
Overseas services	—	4.1	0.1	—	4.2	—	3.9	0.1	—	4.1	—	4.8	0.1	—	4.9
Agriculture, fisheries, and forestry	1.6	14.4	41.8	0.1	58.0	1.7	15.6	42.8	0.2	60.2	1.9	15.1	44.3	—	61.3
Industry, trade, and employment															
Aerospace	—	—	—	0.5	0.5	—	—	—	0.6	0.6	0.0	0.4	0.2	—	0.6
Energy	17.5	59.3	215.8	—	292.6	20.3	67.2	214.8	—	302.3	21.8	67.8	240.4	—	330.0
Other industry, trade and employment	1.4	38.1	9.7	2.9	52.2	1.5	39.2	8.8	3.5	53.0	0.7	21.2	10.2	—	32.1
Environmental services	0.7	25.1	4.2	1.5	31.5	0.8	23.7	3.9	1.5	29.9	0.0	22.3	3.8	—	26.1
Law, order and protective services	0.6	1.7	3.4	—	5.8	0.6	2.0	4.1	—	6.7	—	1.3	4.5	—	5.8
Education and Science															
Agricultural and Food Research Council	27.2	3.2	—	—	30.4	30.8	3.7	—	—	34.5	38.5	—	—	—	38.5
Medical Research Council	30.1	28.9	—	—	59.0	31.7	30.5	—	—	62.2	21.4	45.5	—	—	67.0
Natural Environment Research Council	46.1	10.1	—	—	56.2	47.4	10.4	—	—	57.8	50.9	8.6	—	—	59.5
Science and Engineering Research Council	60.5	—	—	—	60.5	66.4	—	—	—	66.4	70.3	—	—	—	70.3
Other science, education, etc.[b]	8.1	—	—	0.6	8.7	8.7	—	—	0.5	9.2	9.2	—	—	—	9.2
Health and social security	—	8.3	0.3	—	8.5	—	8.5	0.3	—	8.8	—	9.3	0.3	—	9.6
Other public and common services	—	—	0.5	8.8	9.3	—	—	0.7	10.4	11.1	—	—	1.1	—	1.1
Scotland	1.5	6.5	0.5	—	8.5	1.7	7.0	0.5	—	9.2	1.9	7.8	0.6	—	10.2
Wales	—	—	—	—	—	—	0.1	—	—	0.1	—	0.0	—	—	0.0
Northern Ireland	—	5.5	—	—	5.5	—	5.3	—	—	5.3	0.0	5.1	2.1	—	7.2
Total civil	195.3	205.3	276.3	14.5	691.4	211.6	216.9	276.0	16.6	721.1	216.6	209.2	307.5	—	733.2
Total	195.3	381.4	605.6	14.5	1,196.8	211.6	426.1	621.0	16.6	1,275.2	216.6	442.0	641.5	—	1,300.0

Source: Department of Trade and Industry.

Figures exclude R&D in the social sciences and humanities.

[a] Revised.

[b] Includes the Arts and Libraries programme.

Table 1.8 Net central government expenditure on R&D, Britain (£ millions)

Type of expenditure	*1981–82 outturn*	*1982–83 outturn*	*1983–84 outturn*	*1984–85 provisional outturn*
Science and technology R&D				
Defence	1,743.3	1,763.1	1,982.3	2,175.0
Civil				
Overseas services	15.3	17.5	18.9	20.7
Agriculture, fisheries and forestry	111.8	119.2	124.1	131.8
Industry, trade and employment				
Aerospace	173.7	145.9	167.3	178.0
Energy	238.7	244.4	237.7	232.0
Other industry, trade, and employment	117.4	143.4	153.6	195.0
Environmental services	64.8	60.1	59.9	70.9
Law, order, and protective services	8.2	9.2	9.2	10.3
Education and science				
Agricultural and Food Research Council	41.4	43.1	45.2	46.1
Medical Research Council	100.6	106.4	113.0	116.5
Natural Environment Research Council	52.1	55.2	60.0	63.1
Science and Engineering Research Council	212.1	230.0	245.1	268.5
Other science, education, etc.*	14.0	14.8	15.1	15.9
Universities	400.0	440.0	459.0	472.0
Health and social security	18.6	20.3	21.1	21.1
Other public and common services	9.9	11.6	1.3	1.4
Scotland	39.4	41.4	45.5	48.8
Wales	0.5	0.7	0.9	0.9
Northern Ireland	9.2	8.7	13.5	13.8
Total civil	1,627.5	1,712.0	1,790.3	1,906.9
Total science and technology R&D	3,370.7	3,475.1	3,772.6	4,081.9
Social science R&D				
Defence	1.3	1.8	1.8	1.9
Civil				
Overseas services	1.2	1.2	2.1	1.6
Agriculture, fisheries, and forestry	0.1	0.2	0.2	0.3
Industry, trade, and employment	5.0	5.7	7.0	9.5
Environmental services	2.1	2.3	2.4	4.1

Table 1.8 — continued

Type of expenditure	*1981–82 outturn*	*1982–83 outturn*	*1983–84 outturn*	*1984–85 provisional outturn*
Law, order, and protective services	3.0	3.0	3.0	3.1
Education and science				
Economic and Social Research Council	15.3	16.4	17.5	17.6
Other science, education, etc.	8.4	8.9	11.3	13.1
Universities	140.0	160.0	157.0	162.0
Health and social security	9.6	8.6	6.7	6.8
Other public and common services	3.8	3.8	2.8	2.5
Scotland	1.4	1.4	1.4	1.6
Wales	1.1	0.9	0.6	0.8
Northern Ireland	0.5	0.7	0.8	1.0
Total civil	191.5	212.9	212.7	224.0
Total social science R&D	192.8	214.7	214.5	225.9
Total expenditure on R&D	3,563.6	3,689.8	3,987.1	4,307.8

Source: Department of Trade and Industry.

*Includes the Arts and Libraries Programme.

of the total in 1983. Most of the spending on R&D in the social sciences was in universities.

Within the field of science and technology, the biggest single programme was defence, which is almost half the total. About 60 per cent of the expenditure was in private industry. Other important programmes are education and science, which includes the research councils and universities (more than half civil expenditure) and the energy programme. Direct expenditure on research in universities only represents *c.* 11 per cent of the total.

Current spending on intramural research and development in the field of science and technology is analysed in Table 1.7 into basic research, applied research, and experimental development. The largest programme was defence (+4 per cent), where development was the biggest component of current expenditure. Energy was next in importance, where current spending was also heavily concentrated on development. Education and science was the next largest programme, where expenditure was mainly on basic research. Of the total, only 20 per cent was spent on education and science.

Net expenditure is shown in Table 1.8. Expenditure for the

Table 1.9 Net central government expenditure on R&D: analysis by European Community objectives for R&D spending, Britain (£ millions)

Objective	*1982–83*	*1983–84*	*1984–85 (Provisional)*	*1985–86 (Estimates)*
Exploration and exploitation of the earth:				
Total	67.7	69.3	73.8	77.5
General Research	55.4	59.8	63.1	—
Earth's crust and mantle excluding sea bed	2.9	0.4	0.6	—
Atmosphere	5.8	6.1	6.3	—
Other	3.5	3.0	3.9	—
Infrastructure and general planning of land use				
Total	61.1	57.3	62.8	58.6
Construction and planning of buildings	19.5	16.4	17.4	—
Transport systems	23.9	24.1	19.0	—
Civil engineering	8.6	9.2	17.4	—
Other	9.2	7.6	9.1	—
Control of environmental pollution				
Total	49.0	44.0	50.8	53.8
General research	1.2	0.9	1.2	—
Water pollution	4.9	3.7	5.1	—
Atmospheric pollution	3.1	4.5	5.4	—
Radioactive pollution	30.9	31.2	33.9	—
Other scientific research on the pollution of the environment	9.0	3.7	5.1	—
Protection and promotion of human health				
Total	141.7	151.9	156.8	162.2
General research	118.1	132.7	137.4	—
Biomedical engineering medicines	6.7	2.7	2.9	—
Nutrition and food hygiene	6.0	6.8	5.9	—
Other	11.0	9.8	10.6	—
Production, distribution, and rational utilization of energy				
Total	224.2	220.6	218.6	213.0
Nuclear fission	167.8	162.8	155.9	—
Nuclear fusion	23.3	22.1	20.7	—
Renewable energy sources	13.5	12.7	15.3	—
Other research on production, distribution, and rational utilization of energy	19.6	23.0	26.7	—

Table 1.9 — continued

Objective	*1982–83*	*1983–84*	*1984–85 (Provisional)*	*1985–86 (Estimates)*
Agricultural production and technology				
Total	202.0	206.9	214.8	222.4
General research	22.0	12.3	12.5	—
Animal products	58.2	56.4	57.6	—
Fishing and fish farming	16.6	17.9	20.6	—
Crops	74.5	91.4	96.1	—
Other	30.7	28.9	28.1	—
Industrial production and technology				
Total	218.1	257.9	306.9	336.3
General research	26.5	27.7	36.7	—
Manufacturing and processing techniques and materials research	35.8	22.4	29.9	—
Manufacture of motor vehicles and other means of transport	97.8	119.5	129.5	—
Electrical and electronic engineering	18.5	29.4	51.5	—
Other	39.4	58.9	59.3	—
Social structures and relationships				
Total	25.9	33.0	37.5	44.8
Education, training, recurrent education, and retraining	8.8	16.6	22.0	—
Other research with regard to society	17.1	16.4	15.5	—
Exploration and exploitation of space				
Total	73.2	74.1	79.1	66.2
Research financed from general university funds				
Total	600.0	616.0	634.0	651.0
Non-oriented research				
Total	210.6	268.2	291.4	327.5
Mathematics and natural sciences	185.4	193.6	197.5	—
Engineering sciences	3.6	56.0	75.0	—
Social sciences and humanities	20.0	17.8	18.2	—
Other	1.5	0.9	0.7	—

continued overleaf

Table 1.9 — continued

Objective	*1982–83*	*1983–84*	*1984–85 (Provisional)*	*1985–86 (Estimates)*
Other civil research				
Total	58.0	11.1	12.1	12.9
Defence				
Total	1,758.2	1,976.5	2,169.3	2,378.9
Total all objectives	3,689.8	3,987.1	4,307.8	4,605.3

Note: Expenditure under any one objective may be the addition of R&D for more than one programme.

various programmes is given for the four financial years from 1981–82 to 1984–85. R&D in the field of science and technology, as might now be expected, is dominated by spending on defence (53 per cent). Industry, energy, trade and employment, and education and science, including the universities and research councils, are the important civil research programmes. Education and science in all non-defence sectors contributes 28 per cent of the total.

The Statistical Office of the European Communities (SOEC) has, for a number of years, compared the R&D performance of the governments of member states. This analysis concentrates on the 'objective' towards which the spending was directed. As part of a continuing process of improvement, the classification was revised quite substantially by the SOEC during 1983. The statistics in Table 1.9 have been collected on this new basis. The table shows details for the years 1982–83 to 1985–86. For 1985–86 the estimates are shown only at aggregate level, corresponding to what are described by SOEC as the 'chapters' of expenditure. The table includes all expenditure on R&D, in both science and technology and the social sciences and humanities. The imbalances are again quite evident. Whereas 52 per cent of the expenditure goes on defence, only 14 per cent is spent on directly funding research in universities, and 7 per cent on industrial production and technology.

Table 1.10 provides the distribution of persons employed in R&D by the government. They are mainly engaged in work arising from the intramural expenditure shown in Table 1.6. Some of those working in an administrative capacity are, however, managing the extramural projects. Defence and energy are the main fields of employment, reflecting again the high degree of priority allocated to them in Britain.

Table 1.10 Central government employment of R&D, Britain (thousands)

	1 January 1984					1 January 1985				
	Science and technology					Science and technology				
	Scientists and engineers	Technicians	Other[a]	Social science	Total	Scientists and engineers	Technicians	Other[a]	Social science	Total
Defence	4.7	3.9	16.5	0.1	25.3	4.7	3.9	16.6	0.1	25.2
Civil										
Overseas services	0.2	0.1	0.1	0.0	0.3	0.2	0.1	0.1	0.0	0.3
Agriculture, fisheries, and forestry	1.4	0.8	1.2	—	3.4	1.4	0.8	1.1	—	3.4
Industry, trade, and employment										
Aerospace	0.0	—	0.0	—	0.0	0.0	—	0.0	—	0.0
Energy	3.3	1.8	8.2	0.0	13.3	3.1	1.7	8.9	0.0	13.6
Other industry, trade and employment	0.7	0.2	0.6	0.1	1.6	0.7	0.2	0.6	0.1	1.5
Environmental services	0.8	0.0	0.7	0.0	1.5	0.8	0.0	0.7	0.0	1.5
Law, order, and protective services	0.1	0.0	0.0	0.1	0.3	0.1	0.0	0.0	0.1	0.3
Education and science										
Agricultural and Food Research Council	0.6	0.4	0.6	—	1.6	0.8	0.3	0.7	—	1.7
Medical Research Council	1.5	1.1	1.2	—	3.8	1.5	1.1	1.1	—	3.7
Natural Environment Research Council	1.5	0.1	1.2	—	2.8	1.4	0.1	1.2	—	2.7
Science and Engineering Research Council	1.1	0.2	1.5	—	2.9	1.1	0.2	1.5	—	2.8
Economics and Social Research Council	—	—	—	0.2	0.2	—	—	—	0.2	0.2
Other science, education, etc.[b]	0.3	0.1	0.3	0.0	0.7	0.3	0.1	0.3	0.0	0.7
Health and social security[c]	0.2	0.2	0.2	0.1	0.7	0.2	0.2	0.2	0.1	0.7
Other public and common services	0.0	0.0	0.0	0.1	0.1	0.0	0.0	0.0	0.0	0.1
Scotland	0.2	0.1	0.2	0.1	0.6	0.2	0.1	0.3	0.1	0.6
Wales	—	—	0.0	0.0	0.0	—	—	0.0	0.0	0.0
Northern Ireland	0.1	0.2	0.2	0.0	0.5	0.1	0.2	0.2	0.1	0.5
Total employed on R&D in science and technology	16.6	9.4	32.7	—	58.7	16.6	9.0	33.4	—	59.0
Total employed on social science research	0.4	0.0	0.3	0.8	0.8	0.4	0.0	0.2	0.7	0.7
Total	17.1	9.4	33.0	0.8	59.5	17.0	9.0	33.7	0.7	59.6

Source: Department of Trade and Industry.

[a] Including administrative staff. [b] Includes the Arts and Libraries Programme. [c] Excluding hospital staff.

Factors motivating the high civil R&D expenditure in Japan

The expansion in the basic indices of research activity can be explained by the following:

1. vitalization and development of industrial activities;
2. the common will of individuals in the private, academic, and public sectors to promote industrial technology; and
3. the increased supply of engineers and researchers through the infusion from university education, particularly in engineering.

With regard to the first, the increased tempo of industrial activities and the accompanying expansion of the Japanese economy were perceived to require and were the result of a massive increase in industrial R&D. This in turn contributes to the upgrading of the industrial structural base. Through this reciprocal relationship the level of Japan's industrial technology was enhanced, and industrial activities rapidly expanded into advanced and new fields of technology. Government policy concerning industry, education, and the promotion of science and technology is clearly directed towards supporting this industrial growth.

Japanese industry at large has also succeeded in generating an effective strategy for technology promotion. The common will of the private, academic, and public sectors has been harnessed by a well-defined system of national priorities, aggressively promoted. Implementation took the form of pragmatic policies, but resulted in concerted efforts between these three sectors. In industry such policies have been developed in forms tailored to the various stages of technological cultivation.

The first stage involved the introduction of overseas technologies as a leverage to enhance domestic technology and thereby catch up with the advanced countries. The resultant improved technology drives a rigorous strategy for high-quality products. When a certain measure of success is achieved, matters proceed into the second stage. There is a movement from the refinement and improvement of introduced technology into indigenous development, at first observed mainly in aspects of technological applications to conventional products. Such a drive gradually gave expression to the development of original indigenous products.

This industrial technology promotion has now reached the third stage, which involves (a) an emphasis on target-oriented basic research, geared toward the development of technology of a highly original character, (b) the proper evaluation of the results obtained in basic research areas and a more full-scale R&D approach to link these results to applied and developmental research in order to maintain a leading position in the more innovative industrial fields, and (c) a more pronounced inclination towards development of original products.

The attitude of industry toward interaction with academia has also changed greatly in response to these stages of technological development.

During the first stage the initiatives in science and technology depend more on the universities than on industry. Although this tendency has not undergone a basic change even today, in the transition from the second to the third stage there occurred a massive influx of university graduates into industry, so establishing autonomous research systems in the industrial sector itself. Concurrently, the 'catch-up' flavour of the previous stage abated. In certain fields in Japan there grew up a feeling that the advanced countries of the West had been overtaken.

Table 1.11 presents a comparative evaluation of Japanese expertise and that of the west (United States and Europe) in 83 selected technological areas and their respective R&D potential in the future. With regard to conventional technology, Japan is judged to have caught up with or overtaken the western countries in each industrial field at around 1970. With regard to advanced technologies, however, roughly 50 per cent of Japanese opinion considers that there are still many fields in which Japan has not yet overtaken the western countries. This evaluation can be linked to a conviction that R&D starting from the more fundamental technological research must be carried out for a country to keep its competitive edge in the world scene. Indeed, the judgement is now being translated into action.

A report to Prime Minister Nakasone by the Science and Technology Sub-Committee of the Administrative Reform Promotion Council (composed of individuals with administrative experience in industry, academia, and government) recommended that Japan's technological policy in the years ahead must emphasize measures directed towards strengthening basic research. Similarly, MITI has submitted a bill for increased spending on basic research to the National Diet, which was ratified in June 1985.

The trend is similar in industry, when account is taken of the allocation of funds to basic research. The share of basic and applied research costs in relation to total R&D expenditure increased by 1.1 and 3.8 per cent respectively over the last five years. Developmental research expenditure has undergone a corresponding decrease. This emphasis on basic research has resulted in a more positive approach to interaction with other research organizations conducting basic research, most importantly

Table 1.11 Comparative technology levels of Japan, Europe and the United States, as assessed by Japanese expertise

(Current)

Technology level of Japan	A *Lower than USA*	B *Slightly lower than USA*	C *Same as USA*	D *Slightly higher than USA*	E *Higher than USA*
E					
D		80 29 66 56 62 58 67	46 57 71 39 81	76 31	
C		2 40 47 8 43 59 10 5 50 72 26 36 61 25	32 11 35 42 27 54 16 44 45 79 19 51	17	
B	49 60	75 48 20 1 11 37 13 55 64 6 22 73 7 53 69 4 34 63 24 78 3 9 33 52 15	41 68 82 83 38 18 21 65 77 12		30
A	70	14 23 28			

Technology level of Europe

(Future)

Technology level of Japan	A *Lower than USA*	B *Slightly lower than USA*	C *Same as USA*	D *Slightly higher than USA*	E *Higher than USA*
E					
D		26 66 29 5 62	46 42 17		
C	40 47	2 3 23 15 67 48 73 57 10 56 43 9 36 65 72 61 78 52	32 31 16 39 18 58 35 80 19 54 44 81 74 79 63 51 13 27	76	
B	60 20	24 55 7 8 53 69 59 1 38 50 71 4 33 34 64 37 25 77 14 49	45 75 28 83 82 6 21 12 68	30	
A	11	70 22	41		

Technology level of Europe

Source: Science and Technology Agency

Table 1.11 — continued

83 Technological areas

1. Advanced automobile engine parts
2. Advanced structural materials
3. Animal breeding
4. Animal cell cultivation
5. Artificial heart
6. Artificial satellite
7. Assessment of fishery resources
8. Automated translation system
9. Bioadaptable materials
10. Biomass conversion technology
11. CAD/CAM (computer-aided design/manufacture)
12. Coal gasification
13. Coal liquification
14. Commercial aircraft
15. Computer for medical use
16. Construction technique of dam
17. Construction of residential house
18. Cryogenic technology
19. Cutting tool
20. Data base
21. Deep-ocean surveying ship
22. Deep-sea oil production system
23. Development of new drugs
24. Electric wave measurement technology
25. Enhanced air conditioning technique
26. Earthquake forecasting system
27. Environment assessment
28. Exploration of natural resources
29. Facsimile (FAX)
30. Fast breeder reactor
31. Fermentation technique
32. Fire extinguishing technology
33. Food processing technology
34. Food storing and distributing technology
35. Forestation
36. Geothermal electric power generation
37. Gene engineering
38. Global-scale ambient air monitoring system
39. Industrial robot
40. Integrated circuit
41. Laser application for medical use
42. Laser beam printer
43. Light communication system
44. Lumber processing
45. Machine tools
46. Magnetic levitated train
47. Mainframe computer
48. Measurement of air pollutants
49. Microcomputer and personal computer
50. Moving communication technology
51. Multilayer IC (integrated circuit) substrate made of alumina
52. Multipurpose CATV (computer-aided TV)
53. Nuclear fusion
54. Ocean energy utilization
55. Oil sand, oil shale
56. Optical electronics devices
57. Optical fibre
58. Optical fibre with plane of polarization
59. Pattern recognition
60. Phosphoric acid-type fuel cell
61. Physiological study of mental activity
62. Plain paper copying machine
63. Plant breeding
64. Plant seed
65. Plant protection
66. Prevention technology of urban disasters
67. Production of semiconductor
68. Radioactive waste treatment
69. Rocket technology
70. Safety assessment of chemical substances
71. Safety engineering of light water reactor
72. Sensors
73. Solar cell
74. Solar thermal power generation
75. Space communication
76. Traffic control system
77. Uranium enrichment
78. Ultra-high pressure technology
79. Ultra-super vacuum technology
80. Video disk
81. Videotex
82. Weather forecasting
83. Wind energy utilization

Table 1.12 Funding of R&D in Japan in 1983, by sector (¥ billions)

R&D fund sources	*R&D fund*	*%*
Government R&D funds		
National government[a]		
MOE	709	9.9
STA	392	5.5
MITI	178	2.5
MAFF	59	0.8
MHW	29	0.4
Others	89	1.2
Total	1,456	20.3
Local government	268	3.7
Total	1,724	24.0
Private R&D funds		
Industry	4,475	62.3
Private universities	731	10.2
Private research institutes	218	3.0
Others[b]	33	0.5
Total	5,457	76.0
National R&D fund grand total	7,181	100.0

Sources: Prime Minister's Office, MOE, STA, MITI, MAFF, and MHW.

[a] MOE, Ministry of Education; STA, Science and Technology Agency; MITI, Ministry of International Trade and Industry; MAFF, Ministry of Agriculture, Forestry, and Fishery; MHW, Ministry of Health and Welfare.

[b] Includes charity fund from donation foundations and R&D fund from foreign companies for their subsidiaries.

universities. This is especially true in biotechnology and materials research.

In response to the advances in industrial technology and the needs of the industrial sector, Japan's engineering education has also expanded greatly. This is the third element supporting the increasing trend in the basic indices of R&D activity. The number of universities in Japan grew from 245 in FY 1960 to 455 in FY 1982, and the number of junior colleges from 280 to 526 over the same period. The total number of graduates from these respective institutions likewise rose from 120,000 and 30,000 in 1960 to 382,000 and 167,000 in 1983. Of this total, the number of natural science graduates increased from 40,000 in 1960 to 130,000 in 1982, a 3.25-fold increase.

Natural science graduates constitute the pool from which Japan's researchers are mainly drawn, and form the first system (on a human level) of industrial–academic interaction. Industry often has a close relationship with individual professors at the universities from which their researchers graduated, providing a human basis for industrial–academic interchange. As a result, industrial R&D activities have been expanded through the mass influx of engineers and scientists who have received quality education.

The Japanese government believes that it will be very difficult to increase government R&D funding in the coming years despite the recognition of the increasing importance of R&D. Accordingly, it will adopt measures to induce the inflow of private R&D funds into universities and national laboratories and to change the existing distribution shown in Table 1.12. These measures will form the basis of the future policy for industrial–academic interaction.

The decline of the British industrial base

Of the five leading industrial nations, Britain now devotes the smallest share of its GDP to civil R&D. This disadvantage in quality is not offset by any generally greater effectiveness. A combination of different indicators all point to the declining industrial impact of Britain's civil R&D effort. In recent years there have been considerable weaknesses in the supply side of the economy. In particular, information and understanding of key technologies is often inadequate, and the training and education system is not meeting industrial needs to the same extent as found in other leading industrial countries.

Industry must also share the responsibility for our performance. Although the very best British companies are also world leaders with high technological reputations, the overall level of resources that British companies devote to technology is low by international standards. Industrial R&D expenditure has dropped even since 1981, except in electronics, chemicals, and vehicles. But in these areas too our major industrial competitors have increased their expenditure even more than has Britain. Within this declining overall budget there has been greater concentration on short-term product development at the expense of longer-term applied research. There are other disturbing indicators. The British share of US patents has fallen from 20 per cent to 8 per cent since the 1960s, while Japan's has risen from 20 per cent to over 35 per cent. Other countries have exploited the world pool of technology more effectively than has Britain. With over 90 per cent of the world's expenditure on R&D taking place outside Britain, inward technology licensing could play a greater role in innovation. But even to do this effectively requires a strong R&D base.

Japan has transformed its position by ruthlessly exacting a unified R&D strategy, mainly directing this to industrial development, while at the same time investing in the human resource required to support the policy. In Britain the reverse has been the case. An obsessive expenditure on non-productive defence R&D has failed to provide any commensurate benefit in industrial development. Even the effectiveness of this defence R&D must be questioned in view of the failure to achieve major objectives, and lack of control in containing expenditure targets. When more than 50 per cent of the scientific and technological effort is wasted as far as industrial growth is concerned, it is not surprising that the remainder of the effort, spread thinly, has not even maintained Britain's ability to manufacture original products based on the new technologies.

Moreover, the ability to take a leading role in developing new technologies has been weakened because of the lack of priority given to basic research, in universities and in government or industry-funded laboratories. The contrast with Japan in all aspects is stark (Table 1.12). More than double the pro rata resources are devoted to university research, for example. Table 1.9 shows how superficial has been the emphasis and direction. Exploration of the earth's crust (1.6 per cent), planning of land use (1.2 per cent), control of pollution (1.1 per cent), protection of human health (3.1 per cent), and energy research (4.6 per cent) are illustrations of too little money having too little direction. Research into new materials, civil space technology, industrial robotics, cryogenics, biomaterials, biosensors, vacuum technology, and so on are conspicuous only by their absence. Yet Japanese opinion still maintains that British research potential and abilities remain ahead of Japan's (Table 1.11) in many areas of investigation.

In relation to our major concern, namely academic–industrial cooperation, the data provided in this chapter mainly illustrate the poor base on which such interactions depend. Neither public or private sources in Britain are prepared to invest in either basic research or longer-term R&D involving universities or the brightest human products derived from them. Subsequently we must consider how limited resources are utilized in Britain, and see how Japanese experience enables us to evaluate such activities.

Chapter two

Implementation of research and development

Japan

The organizations implementing research in Japan fall into the following broad categories: (1) universities, (2) national and public laboratories, (3) private laboratories, (4) special corporations, (5) industrial research organizations. To examine the nature of the interactive relationship between the first four and industrially related research, the general characteristics of each will be given.

Universities

Japan's university system currently consists of 95 national, 34 public, and 328 private universities, a total of 457. There are an additional 532 junior, national, and private colleges.

The number of universities grew rapidly to accompany the post-war mandatory schooling up to high school level. Upon high school graduation, some 25 per cent of the students advance to universities, 10 per cent to junior colleges, and 10 to 20 per cent to other types of educational institutions. The enthusiasm of the Japanese people for education is a clear manifestation of their ambitions. The government as well has made provision for a complete system of higher education, ensuring equal opportunity and democratization in education after World War II. As of May 1983 there were 1.73 million undergraduates, 43,000 students in master's programmes, and 20,000 in doctoral programmes. The corresponding totals in the natural sciences were 572,000, 28,000, and 13,000, respectively.

The increase in the number of universities has led to an increase in the number of university teachers. At the present time, natural science professors, lecturers, and research assistants in all universities (national, public, and private) number 16,000, 12,000, 9,000, and 29,000, respectively. When other supporting research personnel

Table 2.1 University research in Japan

	University			
	Total	National	Local government	Private
Number of universities	457	95	34	328
Number of R&D personnel (100 persons)	2,430	1,145	162	1,124
Natural sciences total	1,026	582	54	392
Professor	158	79	7	72
Associate professor	122	72	6	44
Lecturer	88	30	6	52
Staff researcher	291	154	19	118
Assistant	87	53	6	28
Technicians	121	93	5	24
Other supports	159	101	5	54
Intramural R&D expenditure (¥ billions)	1,649	711	78	860
Natural sciences total	1,028	561	49	418
Physical sciences	148	129	5	14
Engineering sciences	359	197	13	149
Agricultural sciences	81	55	5	21
Health sciences	441	181	27	234
R&D structure of natural sciences by type of R&D activity (%)				
Physical sciences				
Basic	90	89	99	94
Applied	8	9	1	6
Development	2	2	—	—
Engineering sciences				
Basic	55	50	53	63
Applied	26	25	41	27
Development	19	25	6	10
Agricultural sciences				
Basic	56	55	43	62
Applied	33	33	21	34
Development	11	12	36	4
Health sciences				
Basic	44	40	53	45
Applied	55	59	47	52
Development	1	1	—	3
Total				
Basic	55	56	56	54
Applied	37	33	39	41
Development	8	11	5	5
Sources of R&D funds for natural sciences (¥ billions)				
Government	654	547	49	58
Private	374	14	—	360
Total	1,028	561	49	418

Source: Ministry of Education and the Prime Minister's Office.

and technicians are counted, the total is 103,000. The total number of research employees in universities and junior colleges is 290,000.

Of the ¥1,650 billion in research expenditure for internal use in universities in FY 1983, ¥1,028 billion was directed to natural science R&D. Health sciences (including medicine) accounted for 43 per cent of this total, followed by engineering science, physical sciences, and agricultural sciences at 35 per cent, 14 per cent, and 8 per cent, respectively. In short, the biggest part of the ¥1,028 billion total for FY 1983 natural science R&D funds went to the health sciences (¥441 billion; primarily medical research), ¥359 billion for the engineering sciences, ¥148 billion for the physical sciences, and ¥81 billion for agricultural science. These university R&D funds consisted of ¥904 billion from the central and local government, and ¥745 billion from private R&D funds, giving a total sum of ¥1,650 billion.

Research in the natural sciences can generally be divided into basic, applied, and developmental. Natural science research in universities is mainly basic research conducted in the physical sciences. Although basic research accounts for roughly 55 per cent of the research conducted in the engineering and agricultural sciences, there is also a considerable emphasis on applied and development research, reflecting the practical orientation of these departments.

The emphasis in the health sciences, primarily medicine and pharmacology, is on basic and applied research (see Table 2.1, R&D structure of natural sciences).

National laboratories

National laboratories are technological research organizations set up by the government for the specific policies of the various ministries and agencies. The National Research Institute of Police, for example, is attached to the National Police Agency, which in turn belongs to the Prime Minister's Office. Similarly, the National Institute of Environmental Pollution Research is attached to the Environmental Agency, and the Institute of Public Health to the Ministry of Health and Welfare.

The national laboratories have procedures and provisions for consigned and co-operative research, but as the names of the above institutes suggest, in most instances their nature is such that they cannot readily support corporate research. The possibility of interaction between industry and academia arises only when the corporate policies coincide with the research orientation at these organizations. When this occurs, it is usually in laboratories

Table 2.2 Japanese national laboratories

Code	*Government ministry*	*Agency*	*Number of laboratories*
PMO	Prime Minister's Office	STA	6
		Others	5
MAFF	Agriculture, Forestry, and Fishery		30
MITI	International Trade and Industry	AIST	16
MOE	Education		15
MOHW	Health and Welfare		10
MOT	Transportation		8
MOPT	Post and Telecommunication		1
MOL	Labour		2
MOC	Construction		3
MOHA	Home Affairs		1
Total			97

Source: Prime Minister's Office and Directories.

Table 2.3 R&D in national laboratories, Japan FY 1983

Type of laboratory	*Number of laboratories*	*Number of R&D employees*		*R&D funding (¥ billions)*
		Total	*Researcher*	
Physical	18	3,913	1,466	27
Engineering	38	8,557	4,483	112
Agricultural	29	7,288	3,150	43
Health	12	4,068	1,080	19
Total	97	23,826	10,179	201

Source: Prime Minister's Office and Directories.

Table 2.4 R&D in public laboratories, Japan FY 1983

Type of laboratory	*Number of laboratories*	*Number of R&D employees*		*R&D funding (¥ billions)*
		Total	*Researcher*	
Physical	55	1,790	1,316	13
Engineering	140	4,477	3,348	36
Agricultural	392	16,632	7,683	113
Health	55	2,288	1,611	16
Total	642	25,187	13,958	178

Source: Prime Minister's Office and Directories.

coming under the jurisdiction of the Science and Technology Agency (STA), the Ministry of International Technology and Industry (MITI) and its Agency of Industrial Science and Technology (AIST), or the Ministry of Agriculture, Forestry, and Fisheries (MAFF) (Tables 2.2 to 2.4).

Local governments (on the prefectural and city levels) also have testing and research laboratories to promote industrial technology in their region, as well as to do studies for the public good. All 47 prefectures and the 11 nationally designated and metropolitan centres, which have administrative powers comparable to those of the prefectures, have their own testing and research organizations. Additionally, some cities apart from the 11 nationally designated metropolises have small-scale testing and research centres in order to help local small industry.

Table 2.4 indicates the number of laboratories within this group, the number of researchers employed, and R&D expenditure according to type (basic:applied:developmental), which is in the ratio 29:113:36. There are various aspects in which this structure changes, depending on whether the organization is national or public. The details are presented in Table 2.5.

Special corporations (research organizations)

The term 'special corporations' here designates those nine establishments funded almost entirely by the government, and engaged in specific R&D work (Table 2.6):

1. Power Reactor and Nuclear Fuel Development Corporation (FRNFDC)
2. Japan Atomic Energy Research Institute (JAERI)
3. National Space Development Agency (NASDA)
4. Research Development Corporation of Japan (JRDC)
5. Institute of Physical and Chemical Research (IPCR)
6. Marine Science and Technology Centre (MASTEC)
7. New Energy Development Organization (NEDO)
8. Institute of Agricultural Machinery (IAM)
9. Japan Nuclear Ship Development Agency (JNSDA)

These corporations operate mainly through government funding subsidies and business revenues. Numbers 1 through 6 are attached to the STA, no. 7 to MITI, and no. 8 to MAFF. Number 9 was absorbed into no. 2 on March 31, 1985. In each instance the developmental research is conducted in accordance with government policies for long-term technology development.

Table 2.5 National laboratory research, Japan

	National laboratory		
	Total	*National*	*Local government*
Number of national laboratories	831	115	716
Natural sciences total	739	97	642
Number of research personnel (100 persons)	581	254	327
Natural sciences total	531	238	293
Researcher			
Regular	242	102	140
Non-regular	4	2	2
Assistants	19	11	8
Technicians	80	26	54
Other supporting staff	99	51	48
Intramural R&D expenditure (¥ billions)	400	208	192
Natural sciences total	379	201	178
Physical sciences	40	27	13
Engineering sciences	148	112	36
Agricultural sciences	156	43	113
Health sciences	35	19	16
R&D structure of natural sciences by type of R&D activity (%)			
Physical sciences			
Basic	37	44	22
Applied	44	34	67
Developmental	19	22	11
Engineering sciences			
Basic	14	16	7
Applied	44	43	48
Developmental	42	41	45
Agricultural sciences			
Basic	17	40	9
Applied	62	42	69
Developmental	21	18	22
Health sciences			
Basic	61	66	56
Applied	33	24	43
Developmental	6	10	1
Total			
Basic	22	31	14
Applied	51	39	62
Developmental	27	30	24
Sources of R&D funds for natural sciences (¥ billions)			
Government	378	201	177
Private	1	—	1
Total	379	201	178

Source: Prime Minister's Office.

Table 2.6 Special corporation research, Japan

	Special corporation total
Number of special corporations	13
Natural sciences total	9
Number of research personnel (100 persons)	77
Natural sciences total (100 persons)	74
Researchers	
Regular	25
Non-regular	5
Assistants	14
Technicians	11
Other supports	19
Intramural R&D expenditure (¥ billions)	291
Natural sciences total	288
Physical sciences	212
Engineering sciences	75
Agricultural sciences	1
R&D structure of natural sciences, by type of R&D activity (%)	
Total	
Basic research	3
Applied research	6
Development research	91
Source of R&D funds for natural sciences (¥ billions)	
Government	277
Private	11
Total	288

Source: Prime Minister's Office.

The R&D expenditure of the special corporations totalled ¥558.1 billion in FY1983. Of this total, intramural research expenditure amounted to ¥288 billion, and outside expenditure to ¥83.7 billion.

Over 90 per cent of the research conducted by these corporations is developmental in nature, with contracts given to private corporations to undertake research and trial manufacture of the developed products. The large outside research expenditure of ¥83.7 billion is made to attract specific expertise and technologies not available in-house.

Private laboratories

Private laboratories can be grouped under the following categories: research laboratories in the form of private corporations, research laboratories in the form of foundations (non-profit organizations), and technological research associations. In the first category, private corporations operate their own research laboratories as independent joint-stock corporations. The research results may be transferred to a third party, or a patent license may be granted, so that the laboratory can operate using the revenue from the royalties received.

Those in the second group are private research laboratories with the status of public service corporations operated on a non-profit basis with funds received from donations, fees from supporting members, and revenue from contracted research services. Tax relief is the biggest difference between this group and the first, which results in a different financial management of the laboratories.

Research laboratories which are public service corporations are established by three types of initiatives:

a. where unversities or university professors take the initiative;
b. where industry takes the initiative, with participation by universities or other academia in setting up the laboratory;
c. where the government or a municipality takes the initiative to pave the way for co-operation between industry, the government, and academia with regard to manpower and funding, and support for the laboratory's research actitivities.

Examples of industrial–academic co-operation involving such research laboratories will be introduced in Chapter 3.

The third group is based on the Technological Research Association law, under which private corporations can form a research association to solve common technological problems, with or without government encouragement in the form of loans and/or research contracts. In some instances university professors participate, and the research involves co-operation between industry, academia, and the government. The private corporations are responsible for the costs of personnel working in these technological research associations, and a portion of the general operating expenses. Examples of industrial–academic co-operation in these technology research associations will be given in Chapter 3.

For obvious reasons detailed information about the research funding of these private research organizations is not available. Table 2.7 gives the overall details, and shows the considerable scale of these research institutions. There are currently 295 private natural science research laboratories, employing a total of 26,918 research personnel, of which 1,717 are professional researchers. Expenditure totals ¥399 billion, of which ¥248 billion is intramural expenditure.

Table 2.7 Private laboratory research, Japan

	Private laboratories				
	Total	*P*	*E*	*A*	*H*
Number of private laboratories (natural sciences)	295	40	149	49	57
Number of R&D personnel (100 persons)	269	20	202	13	34
Researcher	172	16	131	8	17
Assistants	40	2	31	2	5
Technicians	25	+	16	1	7
Other supporting staff	33	1	23	2	6
Gross R&D expenditure (¥ billions)	399	73	240	14	71
Intramural R&D expenditure (¥ billions)	248	18	196	9	26
Self-financed	33	3	13	5	12
R&D fund received	243	16	209	4	14
(Intramurally spent)	240	15	186	4	14
R&D fund paid outside	16	+	15	+	+
R&D structure of natural sciences, by type of R&D activity (%)					
Basic research	9	12	3	26	45
Applied research	31	76	25	43	48
Developmental research	60	13	72	31	7
Sources of R&D funds for natural sciences (¥ billions)					
Government	53				
Private	195				
Total	248				

Source: Prime Minister's Office.

Notes: A, agricultural sciences; E, engineering sciences; H, health sciences; P, physical sciences; +, positive but negligible.

The allocation of research funds to basic, applied, and developmental research in the private engineering research organizations (3:25:72) strongly parallels that for private corporations (6:22:72). In contrast, the corresponding distribution for agricultural and physical sciences resembles that of the national laboratories, and the health sciences compare best with the distribution in universities research. This is probably because the type of research is similar in these categories of organizations (see Table 2.7).

Of the total research funding in private laboratories some ¥53 billion comes from research contracts for the government, with the remainder coming from private sources.

Industry

Research activities were conducted by 17,646 corporations in FY1983. The total expenditure was ¥4,560 billion, with ¥4,247 billion of this total invested in intramural work. Research-related employees totalled 412,000, including 225,000 professional researchers.

Table 2.8 Industrial research in Japan

Industry	No. of companies	Intramural R&D expense (¥ billions)	(1)	(2)	No. of persons engaged in R&D (100 persons)	(3)	(4)	R&D structure by type B	A	D	Total (%)
Agriculture, forestry, and fishery	140	6	0.1	0.26	5	3	0.1	19	16	66	100
Mining	29	16	0.3	0.66	13	6	0.3	18	20	62	100
Construction	1,319	101	2.2	0.54	95	57	2.5	6	24	70	100
Manufacturing	16,101	4,257	93.4	2.31	3,943	2,145	95.3	6	22	73	100
Food	1,354	111	2.4	0.70	127	72	3.2	14	24	62	100
Textile	1,128	51	1.1	0.90	68	34	1.5	8	29	63	100
Pulp and paper	322	26	0.6	0.63	32	19	0.8	5	23	72	100
Printing and publishing	53	11	0.3	0.43	9	7	0.3	3	28	69	100
Chemical industry	2,100	775	17.0	3.34	711	378	16.8	12	25	63	100
Drug	438	290	6.4	6.59	214	110	4.9	18	24	58	100
Others	1,662	485	10.6	2.58	497	268	11.9	9	26	65	100
Petroleum and coal products	143	50	1.1	0.26	37	16	0.7	5	31	64	100
Rubber products	164	58	1.3	2.40	58	35	1.5	4	18	78	100
Glass and ceramics	804	113	2.5	1.82	112	54	2.4	7	21	72	100
Steel	152	186	4.1	1.60	129	50	2.2	7	25	68	100
Non-steel metals	236	74	1.6	1.49	59	29	1.3	3	22	75	100
Fabricated metal products	1,949	83	1.8	1.31	88	53	2.3	2	18	81	100
Machinery	2,240	312	6.8	2.57	317	198	8.7	4	18	78	100
Electric machinery	2,096	1,416	31.1	4.70	1,307	787	35.0	4	19	77	100
Electric	776	458	10.0	4.40	390	249	11.1	4	18	79	100
Electronics	1,320	959	21.0	4.85	917	538	24.0	3	20	77	100
Transportation machinery	678	715	15.7	2.66	561	211	9.4	4	22	74	100
Automobile	201	606	13.3	2.89	474	162	7.2	4	21	75	100
Others	477	109	2.4	1.86	87	49	2.2	5	28	67	100
Precision machinery	722	159	3.5	4.02	164	102	4.5	2	23	75	100
Others	1,960	118	2.6	1.40	165	100	4.4	4	19	77	100
Transportation, telecommunication, and public utilities industries	57	180	4.0	1.04	65	41	1.8	4	28	67	100
Grand total	17,646	4,560	100.0	2.03	4,122	2,251	100.0	6	22	72	100

Source: Prime Minister's office.

Notes: (1) composition (per cent); (2) ratio of R&D fund to sales (per cent); (3) number of researchers (100 persons); (4) composition (per cent); B, basic research; A, applied research; D, developmental research.

For the 1,262 corporations, capitalized at ¥1 billion or over, the intramural spending amounted to ¥383.4 billion, some 84 per cent of the total. For the 4,062 corporations capitalized at ¥100 million or over, the corresponding figure was ¥431.2 billion, some 95 per cent of the total.

The ratio of R&D expenditure to sales was highest in the pharmaceutical industry (6.6 per cent), followed by electronics (4.9 per cent), electrical equipment (4.4 per cent), precision machines (4.0 per cent), chemicals (3.3 per cent), transportation machinery (2.9 per cent), and machine equipment manufacturers (2.6 per cent). The corresponding average ratio for the manufacturing industry was 2.3 per cent. The ratio of expenditure in the agriculture, mining, and construction industries was low, with the result that the corresponding average for industry as a whole was 2.0 per cent.

According to industrial sector figures, the big R&D investors were in electrical machinery, which accounted for 31 per cent of the total R&D industrial expenditure, followed by chemicals (17 per cent), transport machinery (16 per cent), general machinery (7 per cent), and steelmaking (4 per cent). The electrical machinery industry here includes the electronics manufacturers (21 per cent of the total), while chemicals includes pharmaceuticals (6 per cent of the total), and transportation machinery includes the automobile manufacturers (13 per cent of the total).

Of the total number of researchers in industry, 35 per cent (or 78,700) are in electrical machinery. The chemicals industry in the broad sense (including pharmaceutical) employed 17 per cent (37,800), of the total.

The structure of the research expenditure according to basic, applied, and developmental research varies greatly depending on the industrial field. For the metal products and machinery in the broad sense (including general machinery, and electrical, transportation, and precision machinery), the ratio of basic research expenditure is low (4 per cent or below), and that of developmental research is 75 per cent or higher. In the agricultural, mining, food, chemicals, textiles, ceramics, iron, and other materials and related industries, however, basic research accounts for 7 to 19 per cent of the expenditure, while the ratio of developmental research ranges from 58 to 72 per cent.

As will be demonstrated later, the extent of need for basic research shows a definite correlation with the level at which industrial–academic interaction occurs (see Table 2.8).

Britain

The organizations implementing research in Britain fall into the following broad categories: universities and other institutions of higher education, research council establishments, government establishments and laboratories, industry, and independent and grant-aided bodies. These organizations differ greatly in size, structure, funding, and approach.

Universities

The government department responsible for policy regarding the universities and for promoting much of Britain's civil scientific R&D is the Department of Education and Science (DES). The DES is responsible for the allocation of the 'Science Budget' to the University Grants Committee (UGC), the Research Councils, British Museum and Royal Society, and is also concerned with government science policy and international science co-operation. Relations with the universities are conducted through the UGC, and its responsibilities for civil science and technology are discharged mainly through the Research Councils. The UGC acts as a link between the government, from which it receives a block grant, and the universities to which it allocates this grant.

In 1983–84 51 universities were on the Universities Grants Committee list for Great Britain, which includes two business schools and counting separately the seven constituent institutions of the University of Wales. There are two universities in Northern Ireland. In 1983–84 there were 31 polytechnics. Universities and polytechnics exist for scholarship and research, in all disciplines, as well as for teaching. Nevertheless, it is the universities that have, and will retain, the chief role within higher education in basic or fundamental research. On the other hand, universities, polytechnics, and other institutions undertake applied research directed towards industry, commerce, and the public services. All institutions may apply for research council money for specific projects; for the universities alone the 'dual support system' provides an element for research within the block grants made by the UGC. As a result of an evaluation of the effectiveness of such research, from 1986 the funding will be disproportionate within and between universities, depending on performance. This is a very new and controversial principle.

Tables 2.9 through 2.11 set out some basic statistics concerning British higher education and are likely to change markedly in the forseeable future.

Table 2.9 UK higher education institutions, 1983–84

Type of institution	*Number of Institutions*[a]	*Full-time sandwich HE students*[a] *(000s)*	*Part-time HE students*[a] *(000s)*	*Total HE students*[a] *(000s)*
Universities	53[b]	301	36	337
Polytechnics[c]	31	157	71	228
Scottish central institutions	14	14	1	15
Other colleges (maintained)	411	79	125	203
Direct grant/voluntary[d]	56	30	7	38
Open University[e]	1	—	76	76
Total	566	581	316	897

Source: Department of Education and Science.

[a] Home and overseas students, all levels of study.

[b] There are 51 university institutions on the UGC grant list for Great Britain, which includes two business schools and separately the 7 component institutions of the University of Wales. There are two universities in Northern Ireland.

[c] Includes the Ulster Polytechnic, which subsequently merged with the New University of Ulster on 1 October 1984.

[d] Includes Rolls-Royce Technical College, which closed in 1984, and St. Mary's College, Fenham, which was closed in the summer of 1985, but excludes the Cambridge Institute.

[e] Open University data are provisional and relate to the calendar year 1984.

Table 2.10 UK higher education. Home full-time and sandwich students, analysed by sector and level of study

Sector and level of study	*1979–80*	*1980–81*	*1981–82*	*1982–83*	*1983–84*
Universities					
Postgraduates	31	32	32	30	31
1st Degree	232	240	243	241	235
Other advanced	2	2	2	2	2
Total	265	273	276	273	268
Public sector higher education (PSHE)					
Postgraduates	12	12	12	11	11
1st Degree	99	125	142	157	169
Other advanced	91	71	75	81	86
Total	202	208	229	249	266

Source: Department of Education and Science.

Figures indicate thousands of students. Some of the figures may not add up precisely owing to roundings.

Table 2.11 UK higher education. Home full-time and sandwich students, analysed by subjects of study — universities

Subject of study	*1979–80*	*1980–81*	*1981–82*	*1982–83*	*1983–84*
Education	11	11	11	10	10
Medicine	29	30	30	30	30
Engineering	32	34	35	35	35
Agriculture	6	6	6	6	5
Science	60	63	65	66	65
Social studies	63	65	65	63	62
Professional and vocational subjects	5	5	5	5	5
Languages	33	34	35	34	33
Arts	24	24	25	24	22
Total	265	273	276	273	268

Source: Department of Education and Science.

Figures indicate thousands of students.

Research Councils

There are five research councils:

Science and Engineering (SERC)
Agriculture and Food (AFRC)
Medical (MRC)
Natural Environment (NERC)
Economic and Social (ESRC)

In addition to conducting research in their own research establishments and units, the councils support research through: postgraduate awards at universities and colleges in the form of studentships, and fellowships; provision of grants for specific research projects in universities and polytechnics; and provision of grants for research in other research institutions. The ESRC is not relevant to this book.

Science and Engineering Research Council

The SERC operates and supports the following research laboratories and programmes: Rutherford Appleton Laboratory, Oxfordshire; Daresbury Laboratory, Warrington; Royal Greenwich Observatory, East Sussex; and Royal Observatory, Edinburgh. The SERC also contributes to and collaborates with the European Organization for Nuclear Research (CERN), Deutsches Elektronen-Synchrontron (DESY), the Anglo-Australian Telescope, the UK Infra-Red Telescope (Hawain), the European Science Foundation, and the NATO Science and Collaborative Research Programme. SERC is organized on the basis of four boards: Astronomy, Space and Radio; Nuclear Physics; Engineering; and Science. Table 2.12 shows details of the subjects covered in the SERC programmes and the expenditure in cash terms for the years 1983–84 to 1987–88. Table 2.13 shows the distribution of funding among different classes of recipient.

SERC has implemented programmes to bring universities and industry closer together in research in science, engineering, and applied problems of national importance. These schemes are detailed in Chapter 3.

Agricultural and Food Research Council

The AFRC operates the following research establishments: the Animal Breeding Research Organization, the Letcombe Laboratory, the Weed Research Organization, the Poultry Research Centre, the Institutes of Animal Physiology, and Research on Animal Diseases. It also operates 22 state-aided institutes, including the East Malling Research Station, the Long Ashton Research Station, the Rothamsted Experimental Station, and the Animal Virus Research Institute. It also promotes international collaboration. The primary aims of the AFRC are to advance scientific knowledge relevant to agriculture, horticulture, and food and to apply this knowledge to increasing the efficiency of production and utilization of farm products, without detriment to the environment or to the welfare of farm animals.

Medical Research Council

The MRC operates and supports the following research establishments: National Institute for Medical Research, Mill Hill; and Clinical Research Centre, Harrow. It also operates 57 research units, including the Laboratory of Molecular Biology (Cambridge), the Cyclotron Unit (London and Edinburgh), Cell Biophysics (Kings London), Epidemiological Studies in Psychiatry (Edinburgh), and Radiobiology (Harwell). The primary objective of the MRC is to advance knowledge that will improve the health of individuals in the community; this objective requires work on social, environmental, and preventive aspects of medicine as well as studies directly related to the cause, diagnosis, and treatment of disease. The council also aims, with the universities, to maintain and develop the scientific potential of the country in order to encourage a balanced development of new knowledge ranging from fundamental research to practical problems of medicine.

The council seeks to ensure that discoveries made with public

Table 2.12 Science and Engineering Research Council — subject area summary (£ millions)

Subject area	*1983–84*	*1984–85*	*1985–86*	*1986–87*	*1987–88*
Engineering Board					
Environment	1.3	1.3	1.1	1.1	1.1
	3.0	4.1	4.3	4.6	4.9
Machines and power	2.4	1.9	1.9	2.2	2.4
	5.4	6.0	6.9	6.0	6.1
Materials	3.3	2.3	2.2	2.4	2.6
	5.4	5.4	5.9	4.6	4.8
Energy	0.3	0.3	—	—	—
	0.7	0.3	—	—	—
Polymer engineering	—	—	—	—	—
	1.9	1.6	—	—	—
Teaching company	—	—	—	—	—
	1.7	2.3	2.8	3.3	3.7
Biotechnology	—	—	—	—	—
	1.1	1.6	2.0	2.4	4.6
Information technology	—	—	—	—	—
	19.5	28.9	33.8	33.5	34.5
Application of computers to manufacturing engineering	—	0.5	0.6	0.6	0.8
(ACME)	—	3.2	3.7	5.1	5.1
Total technology	0.5	0.5	0.5	0.5	0.6
	—	—	—	—	—
Process engineering	—	0.9	1.7	1.8	2.0
	—	1.9	4.4	4.4	4.1
ESRC/SERC	—	1.1	1.0	0.8	0.9
	—	—	—	—	—
Engineering processes	0.8	—	—	—	—
	5.5	—	—	—	—
Establishments and	—	—	—	—	—
domestic facilities	6.2	5.1	5.3	4.0	1.4
Marine technology	6.3	5.9	5.8	5.5	5.2
Total Engineering Board	65.2	75.1	85.9	82.8	84.8
Nuclear Physics Board					
Nuclear structure	1.9	2.1	1.9	2.2	2.3
Particle physics	4.8	5.7	5.5	5.8	6.0
Establishments	21.8	21.1	20.9	22.5	42.7
International subscriptions	32.1	35.7	38.0	38.0	18.3
Total Nuclear Physics Board	60.6	64.6	66.3	68.5	69.3
Astronomy Space and Radio Board					
Astronomy	6.5	6.2	7.8	7.0	8.0
	0.1	0.2	0.2	0.2	0.3
Solar systems	3.5	3.4	4.3	3.8	4.4
Establishments	23.5	25.2	21.8	21.9	21.0
Other domestic facilities	−1.0	0.1	0.5	0.7	−0.1
International subscriptions	11.6	13.4	16.0	17.1	18.6
Total Astronomy Space and Radio Board	44.2	48.5	50.6	50.7	52.2

Table 2.12 — continued

Subject area	*1983–84*	*1984–85*	*1985–86*	*1986–87*	*1987–88*
Science Board					
Biology	14.6	15.6	16.5	17.0	18.2
Chemistry	14.2	15.7	16.7	17.3	18.4
Maths	3.0	3.2	3.4	3.4	3.6
Physics	7.0	7.5	8.3	8.6	9.3
Other research	3.5	5.2	6.6	6.4	6.6
Establishments	9.2	24.1	25.0	29.4	27.6
Other domestic facilities	1.0	0.6	0.5	0.5	0.5
International subscriptions	20.8	7.1	7.2	6.3	7.2
Total Science Board	73.3	79.0	84.2	88.9	91.4
Central Support	1.4	1.6	1.8	1.7	1.7
Total	245.1	268.5	286.8	292.9	299.2

Source: Science and Engineering Research Council.

Columns may not sum to totals shown owing to rounding.

Table 2.13 Distribution of funding of UK research by SERC (£ millions)

Sector	*1983–84*	*1984–85*	*1985–86*	*1986–87*	*1987–88*
Intramural	92.1	93.9	97.3	98.0	100.3
Research councils	—	—	—	—	—
Other government departments	1.3	0.6	0.6	0.7	0.7
Universities	78.8				
Private industry	0.1				
Research associations	0.1				
Overseas	52.6				
Non-industrial research institutes	0.1	185.3	203.0	209.2	213.2
Professional and learned societies	0.1				
Persons	29.9				
Others	0.1				
Less receipts	10.0	11.4	14.1	15.0	15.0
Total	245.1	268.5	286.8	292.9	299.2

Source: Annual review of government-funded R&D.

money contribute to the nation's economic performance. The council hence aims to secure the successful transfer to British industry of inventions, techniques, and knowledge developed by MRC-supported research teams. It currently has about 100 inventions licensed or under exploitation by the British Technology

Group and some 50 exploitation agreements made directly with commercial organizations.

The council also plans to set up a Centre for Collaborative Research to promote the commercial exploitation of MRC (and university) research by British industry.

Natural Environment Research Council

The NERC operates or supports the following research establishments: the Institute of Geological Sciences, the Institute of Hydrology, the Institute of Marine Biochemistry, the Institute of Oceanographic Sciences, the Institute for Marine Environmental Research, the British Antarctic Survey, and the Freshwater Biological Association. NERC also contributes funding to collaborative programmes with the International Phase of Ocean Drilling (IPOD), UNESCO; NATO; and the OECD.

The NERC is responsible for encouraging, planning, and executing research in those sciences, physical and biological, that relate to humanity's natural environment and its resources. Such investigations seek to provide a better understanding not only of the nature and processes of the environment in which we live and on whose resources we depend, but also of their influence on humanity's activities and welfare and, of growing importance today, of our influence on the environment.

The NERC has identified areas of high priority on which it is concentrating available resources. These include aspects of deep geology, processes in shallow seas and the shelf edges, environmental biotechnology, acidification processes, atmospheric chemistry, climatology, land use, and applications of remote sensing. Many of these initiatives are pursued in collaboration with universities, industry, customer departments, and other research organizations.

Government research establishments

Many government departments operate their own research establishments for the specific use of the various departments. These departments are the Ministry of Defence (MOD); the Department of Energy (DOE); the Department of Trade and Industry (DTI); and the Ministry of Agriculture, Food, and Fisheries (MAFF). Other departments, such as the Home Office and the Department of Transport, operate their own research establishments, but their expenditure is relatively small compared with that of the above four. Some of the major establishments are detailed below.

Ministry of Defence

The MOD operates the following research establishments: the Aeroplanes and Armament Experimental Establishment, Admiralty Marine Technology, the Admiralty Surface Weapons Establishment, the Atomic Weapons Research Establishment, the Chemical Defence Establishment, the Military Vehicles and Engineering Establishment, the National Gas Turbine Establishment, and the Explosives and Rocket Motor Establishment.

Department of Energy

The DOE administers policy development in atomic energy, coal, oil, gas, and electricity. A significant part of the department's R&D expenditure goes to the United Kingdom Atomic Energy Authority (UKAEA). The UKAEA operates the following establishments: the Atomic Energy Research Establishment, Harwell; the Winfrith Atomic Energy Establishment; the Culham Laboratory, Abingdon; the Risley Nuclear Power Development Establishment, Warrington; the Springfields NPDE, Preston; Dounreay NPDE; and Sellafield NPD Laboratories, Cumbria.

Department of Trade and Industry

The DTI operates the following laboratories: the National Physical Laboratory, Teddington; the National Engineering Laboratory, Glasgow; the Warren Spring Laboratory, Stevenage; the Laboratory of the Government Chemist, London; the Computer Aided Design Centre, Cambridge. The DTI is responsible for many of the policies directed at improving research links between universities and industry; these are detailed in Chapter 3.

Ministry of Agriculture, Fisheries, and Food

The MAFF operates the following research establishments: the Royal Botanic Gardens, Kew; the Central Veterinary Laboratories and 23 regional veterinary investigation centres; 12 regional agricultural science laboratories; and 19 experimental husbandry farms.

Industry

Nationalized industries

A number of nationalized industries of central importance to the country's economy are run by public corporations and organize their own R&D activity, for example, the British Steel Corporation, the National Coal Board, and the electricity supply industry. The

Figure 2.1 Expenditure on R&D by the nationalized industries, 1985–86

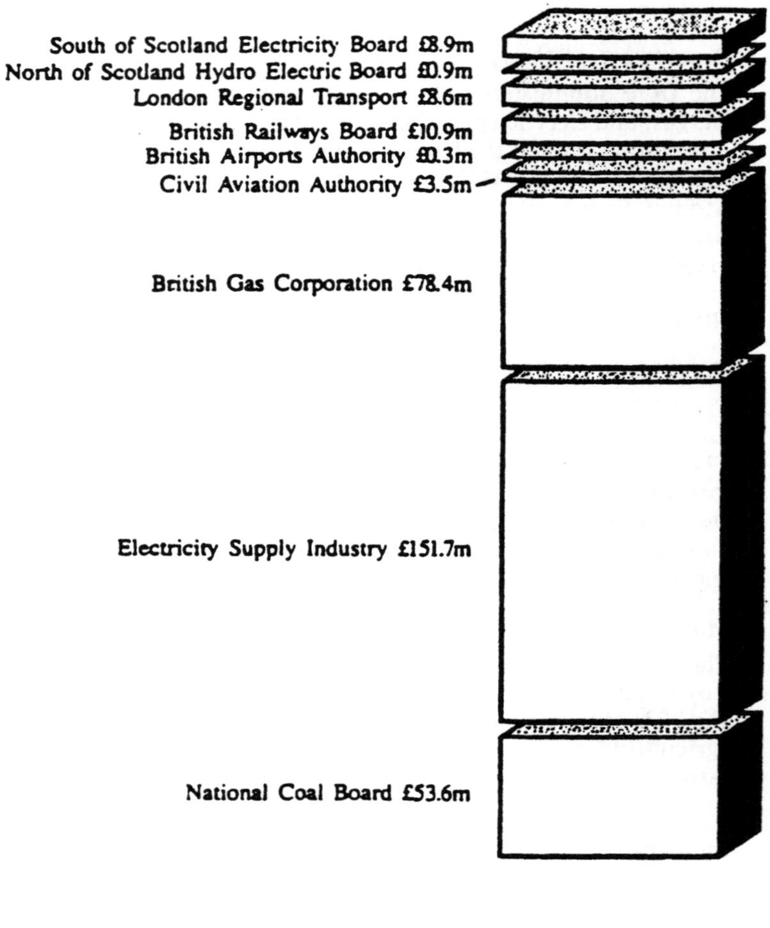

Table 2.14 Industrially performed R&D in Britain at current and constant (1975) prices (£ millions)

Product	*1966*	*1967*	*1968*	*1969*	*1972*	*1975*	*1978*	*1981**	*1983*
			At current prices						
All product groups	580.4	604.5	638.7	680.3	830.5	1,340.1	2,324.3	3,792.5	4,163
All manufactured products, including:	570.2	594.2	629.5	670.6	806.2	1,293.4	2,243.3	3,549.0	3,900
Chemical and allied products	84.3	86.1	93.9	104.5	137.8	245.4	424.7	654.7	765
Mechanical engineering	59.7	61.8	63.1	60.3	54.8	103.4	181.1	248.1	260
Electronics	111.6	114.7	127.9	141.1	179.0	279.0	649.6	1,153.3	1,390
Other electrical engineering	37.0	37.6	39.7	39.5	40.1	73.0	100.8	120.8	117
Motor vehicles and parts	45.2	44.7	44.4	48.9	58.1	88.3	129.7	180.4	239
Aerospace equipment, manufacturing, and repairing	129.0	143.2	150.2	157.8	208.3	291.5	424.6	762.9	720
Other manufacturing	103.4	106.1	110.3	118.4	127.9	212.8	332.6	428.9	406
			At constant 1975 prices						
All product groups	1,504.4	1,514.1	1,506.3	1,512.9	1,417.6	1,340.1	1,566.2	1,661.2	1,564
All manufactured products, including:	1,477.9	1,487.8	1,484.6	1,491.4	1,376.2	1,293.4	1,511.6	1,554.5	1,465
Chemical and allied products	218.3	216.1	233.3	236.5	230.8	245.4	283.5	277.1	278
Mechanical engineering	154.8	154.5	148.0	132.9	93.7	103.4	118.3	111.2	98
Electronics	287.6	285.1	296.8	308.5	301.5	279.0	441.9	510.6	528
Other electrical engineering	93.3	91.8	91.4	85.0	67.9	73.0	68.6	53.4	44
Motor vehicles and parts	119.5	112.8	105.9	111.6	96.9	88.3	87.8	80.3	91
Aerospace equipment, manufacturing, and repairing	330.9	355.8	354.4	350.9	366.4	291.5	285.4	336.7	272
Other manufacturing	273.5	271.6	264.7	266.0	218.9	212.8	226.1	185.1	151

Source: Department of Trade and Industry.

Note: Industry is grouped according to the 1968 Standard Industrial Classification.

*Revised.

organization of research in these industries is the responsibility of the management of a board/corporation/council, but overall R&D programmes are agreed in consultation with the relevant government department (either the Department of Energy or the Department of Trade and Industry). The board/corporation carries out its R&D programmes in its own research establishments or arranges for them to be effectively carried out on its behalf. The research establishments are, therefore, not a direct part of the government organization. The nationalized industries may sponsor research directly in the universities. Figure 2.1 shows the expenditure on R&D.

Private industry

At 1975 prices, £97m less was spent on R&D in industry in 1983 than in 1981 (*British Business*, January 1985). Total amounts of R&D from expenditure are not available, but it is estimated that R&D costs had risen by about 16.5 per cent between 1981 and 1983, and that the decline in volume was, therefore, about 6 per cent over the two years. Table 2.14 shows the detail for the total expenditure on seven groups of manufactured products. The pattern of change was not uniform across industry. The volume of R&D of the chemical, electronics, and motor vehicle industries was estimated to have been greater in 1983 than in 1981, and that in mechanical engineering, other vehicle engineering, aerospace, and other manufactured products declined, in some cases to levels lower than in 1979. The fall in aerospace expenditure was part of a pattern that may be traced back to the peak levels of the 1960s, and expenditure in mechanical engineering R&D fell by 11 per cent in volume between 1981 and 1983. Expenditure on electrical engineering, comprising basic electrical equipment and insulated wires and cables, was estimated to be 17 per cent lower than the 1981 level and 36 per cent lower than in 1978.

Independent and grant-aided bodies

Technology centres (CDRA — Federation of Technology Centres)

Technology centres carry out industrial R&D under contract to industry and to government, and provide consultancy services and product testing for industry. They can act as a route for new ideas from university or government research laboratories into industry. Technology centres fall into six main groups: industrial and domestic materials, building and construction industries, industrial and domestic products, engineering industries, textile industries, and technological services to industry. Its combined annual turnover in 1983 was over £80 million, financed by industry and government on a contract basis.

Independent Contract Research Organizations

The ICRO acts as a trade association, and its twelve members are financed solely by outside research contracts. The twelve members are: the Electrical Research Association, the Fulmer Research Institute, the Huntingdon Research Centre, International R&D, Inveresk Research International, Ricardo Consulting Engineers, Robertson Research International, the Motor Industry Research Association, the British Cast Iron Research Association, the Welding Institute Research Laboratory, the British Nuclear Fuels Metals Technology Centre, and BHRA Fluid Engineering.

The Royal Society

The Royal Society's *raison d'être* from its founding in 1660 has been the promotion of the natural sciences, including mathematics and all applied aspects such as engineering and medicine. It functions in a way similar to international academies overseas in encouraging both national and international scientific activities. Monies are expended to provide services in support of the British scientific community's research activities. These services take the form, for example, of research appointments, exchange fellowships, and travel grants; support of scientific publications and discussion meetings; assistance to other learned bodies in science; some small provision of grants for research, and support of British involvement in international scientific programmes and scientific affairs generally. The Society itself also periodically engages directly in research projects with overseas partners.

British Council

The British Council promotes international collaboration in research. It sponsors visits and exchanges and is supported mainly by the Foreign Office.

Technical Change Centre

The Technical Change Centre was established in 1980 with the support of the Leverhulme Trust and the SERC to develop a major programme of research on the choice, management, and acceptability of technical change relevant to the advancement of the national economy.

Charitable foundations

Includes such famous charities as the Imperial Cancer Research Fund, the Nuffield Foundation, and the Wolfson Foundation.

Japan and Britain

A qualitative assessment would indicate that the R&D infrastructure in Britain approaches that in Japan. The broad pattern of R&D distribution involves universities, national laboratories in basic reseach, and those national establishments directed at particular industrial sectors. As noted in Chapter 1, the investment in basic research is at a considerably lower level in Britain than in Japan, and additionally when defence is excluded the proportion spent on productive industry approaches half of the Japan investment.

There is a sixfold greater proportion invested in university research in Japan than in Britain. Industry in its productive aspects invests almost three times more than in Britain. Our nationalized industry and SERC establishments give nearly the same support as that which is given to national and special co-operations in Japan. Again, the R&D defence expenditure exerts a vast distorting influence on those funds which can be allocated to speculative research upon which our future industrial performance may ultimately depend.

The major divergence in academic research funding is a reflection of the difference of national consciousness in the two countries. Research structures are uniform and built into the Japanese system, whereas in Britain they are haphazard and non-homogeneous. They are directed at training and wealth creation in Japan, whereas in Britain in many instances they operate as a personal staff development activity. There are, however, centres of excellence. We now consider how this British system is being harnessed for industry, relative to similar efforts in Japan.

Part II

Systems of Academic–Industrial Interaction

Chapter three

Industry–academic interaction: systems and government policy

Japan

Industry–academic interaction policies at national and local government levels are of two types: first, those instituted within the various ministries and agencies and second, those government procedures operating through budgetary measures for the implementation of scientific and technological policy.

MOE industry–academic interaction policy

MOE industry–academic interaction policies are of two types: those operated with national universities, which come under its jurisdiction, and those operating through the MOE special corporation, the Japan Society for the Promotion of Science. Those programmes associated with national universities are: acceptance of contract research, promotion of collaborative research, systems for accommodating research personnel on contract from industry, a scholarship grant system, and the Japan Society for the Promotion of Science.

Contract research

Outline Since teachers in Japan's national universities have the status of national civil servants, with their salaries and research expenditure provided by the national budget, it is ruled that the acceptance of contract research from an external sponsor is subject to the approval of the head of the university or the authorized head of the department.

Once approved, the contract is carried out as public business, but the costs needed to implement the research are paid by the contractor and paid into the National Treasury. As a precondition, the research in question must be significant within the context of educational research being undertaken at national universities, and at the same time must present no barriers to the freedom of educational research. There are, in addition, the following terms imposed upon acceptance:

1. Patent rights arising from the research results are the property of the government. The sponsor or the party designated by the contractor can obtain the rights to use the patent on a priority basis for a period of up to 7 years after the completion of the research in question, but a fee must be paid to the government for the use of the contract.
2. Facilities and equipment purchased with the fees will not revert to the contractor.
3. The contract fees must be paid before the initiation of the research, and the contractor cannot order the research to be halted without the consent of the university. Nevertheless, the research can be suspended or its term extended in specific circumstances. Here also the government does not accept responsibility, and as a general rule the fees paid do not revert to the contractor.

In addition to direct expenses, surcharges at approximately 27 per cent for indirect expenses are demanded for research not having a public welfare character. Such research occupies about 60 per cent of the total.

Note 1. The restrictions within the contract research systems in laboratories of other ministries and agencies is virtually the same as for the MOE, but the extent of such contracts is much less. In FY 1983, for example, of the 16 laboratories of the AIST thought to have a strong connection with industry due to the nature of the research they conduct, only one such contract is recorded.

Note 2. Private universities are not bound by the MOE stipulations, and have set up their own systems for the receipt of research contracts.

The current state of the operation of the system for receipt of research consignments National universities receive contracts of research not only from private corporations, but also from other sources, such as the national government, local governments, and special corporations. Of the 1,324 contracts received by national universities under MOE jurisdiction (totalling ¥2,369 million) in FY 1983, 241 were derived from private corporations. This is equivalent to 18 per cent of the total number of such contracts, and accounts for about 13 per cent of the total on a monetary basis (Table 3.1).

Approximately 80 per cent of such contracts derive from either national and public laboratories, or special corporations that serve

Table 3.1 Contract research in national universities, Japan

Source	*Contracts*	*Research (%)*
Central government	428	22
Local government	299	23
Special corporation	313	24
Industry	241	18
Other private organizations	43	3
Total	1,324	100
Total consigned value	¥2,369 million	

Source: Ministry of Education.

Table 3.2 Assessment of industrial–academic interaction by private corporations

Type of interaction	*Number of companies*	*%*	*Assessment by the company (average score)**
Dispatching researchers to university	276	27	3.5
Technical guidance by university	273	26	3.5
Contract research	154	15	3.0
Use of university's facilities and/or equipment	126	12	3.6
University speakers	121	12	3.2
Co-operative research	86	8	3.1
Total	1,036	100	—

Source: Science and Technology Agency survey.

*The assessment is made according to the following rating system: 5, very effective; 3, effective; 1, poor; 0, don't know.

to promote scientific and technological policies on behalf of the national government. The research in question is either basic or applied research of a highly 'public welfare' nature.

Such industry–academic interactions are diverse and include seconding personnel from corporations, advice or consultation furnished by university professors, the use of facilities and equipment, invitations to speakers, and collaborative research. According to the results of a questionnaire distributed to 534 corporations (see Table 3.2), 15 per cent make use of the contract research system, making this the third most popular form of interaction, after the seconding of research personnel from the company and guidance from university professors.

Collaborative research

Outline In May 1983 the MOE issued an edict entitled 'Collaborative Research with the Private Sector and Others', in which it inaugurated the system for collaborative research efforts by university researchers and those in the private sector on themes of common interest. As a general rule, the proportion of the burden of such research costs is 2 to 1, (private to national). In the event the collaborating university must pay ¥360,000 in fees to the National Treasury to defray some of the operating costs of university facilities.

The ratio of ownership rights to patents arising as a result of research collaboration is determined by mutual agreement between the university and the corporations, in accordance with the extent of their respective contributions. Rights to use these patents are given to the partner on a priority basis for the five years immediately following the completion of the research. If the private partner corporation in the collaboration decides, without proper cause, not to use these rights, the university may furnish these to a third party.

The current state of the collaborative research system Only a relatively short time has elapsed since the inauguration of the collaborative research system. In FY 1983 21 national universities started collaborative research with more than 26 private corporations on 56 different themes. In FY 1984 collaborative research was conducted on 160 themes, with 184 researchers seconded from corporations to universities. The use of this system continues to expand rapidly.

Although this is partly due to the fact that the sharing of patent rights makes this form of research more attractive to corporations than contract research, it is also thought that collaborative research affords a greater exchange of information and degree of mutual scientific and technological benefit owing to the discussion and co-operation involved, as opposed to the one-way dependence on universities in the case of contract research. For this very reason there is a need for an ability to collaborate on the part of the corporation, and thus there is a tendency for large companies to participate. It is also thought that the use of this system will increase even further if the operation is made more adaptable. The government has noted this need, and increased the funding for the collaborative research system in the budget to ¥751 million and ¥1,285 million for FY 1984 and 1985, respectively, as compared with the corresponding ¥350 million figure for FY 1983.

Acceptance of company personnel on secondment to universities

Outline This system was set up for the secondment of engineers and scientists, from private corporations, national ministeries and agencies, local governments, and special corporations, for training at the postgraduate level at universities. Corporations rate this system highly, and make extensive use of it.

As can be seen in Table 3.2, almost 30 per cent of the corporations surveyed in the questionnaire dispatch research personnel to universities, and rate highly the effectiveness of the training they receive. This is no doubt due to the fact that the corporation can benefit from the opportunities for contact with university expertise in the course of actual research guidance and discussions.

Operation of the secondment system According to MOE data, the quota for research personnel secondment has been rising year after year, but the requests from corporations exceed this quota (Table 3.3).

Table 3.3 Research personnel contract system, Japan

	1975	*1981*	*1983*
Planned consignments	645	715	814
Actual consignments	651	733	833
Source			
Private industry	—	—	707
MITI and STA	—	—	67
Special corporations	—	—	30
Local governments	—	—	24
National budget for research personnel consignments (¥ millions)	92	103	177

Source: Ministry of Education.
Figures represent individual persons.

A prescribed fee is paid to the National Treasury for such secondments. This is to defray some of the operating costs of maintaining the facilities, and is thus fairly low, in the order of ¥270,000 per person. Payment of this fee entitles the engineer/scientist to study at the postgraduate level.

Note 1. Universities also have an industrial student system, allowing qualified individuals to receive research guidance from faculty members, with the same status as university students. This requires both a written pledge by the candidate that the research is personal and a written statement by the head of the candidate's department that the research is not for corporate aims.

Scholarship grant system

Outline The scholarship grant system provides for official donations to universities by industry in the form of scholarships to encourage research. Normally the grant is made to a specific professor or researcher and is deposited in the National Treasury. After a certain amount is deducted, the professor or researcher can make use of the grant within the context of his or her education and research. There is considerable freedom in the use of this grant; and as far as the recipient is concerned, no official obligations are attached to it, since it takes the form of a donation. In practice, however, industry gains some benefits.

The ownership of patent rights achieved by professors or researchers in national universities as a result of their status as civil servants can be classified into two types, according to the origin of the research funds involved:

(i) When the research is implemented through special research funds or equipment provided by national programmes, the government inherits the patent rights.

Patent rights arising from collaborative research with the private sector are shared with the partner in the event of joint invention, whereas the government inherits the rights in the event of independent invention by the university researcher. In the latter instance, the partner in the collaboration or the party designated can use these rights or a priority basis for a period of 5 years.

Patent rights arising from contract research are the property of the government, and the contractor can make use of these rights on a priority basis for a period of 7 years.

(ii) Patent rights arising from research not of the above nature revert to the individual faculty member who is the inventor. Patent rights arising from research efforts funded by scholarship grants revert to the individual who is the inventor.

For this reason, when an industry–academic co-operative research effort is conducted on specific themes through the supply of R&D funds in the form of a scholarship or grant, the patent disposal is at the discretion of the professor. Ordinarily, patent rights for such inventions are transferred to the corporation without compensation. However, if the invention proves of great benefit to the corporation, there is an unwritten understanding that the corporation side will furnish a sizeable return to the professor.

National university regulations prohibit the receipt of scholarships or grants which specify that patent rights are to be transferred to the contractor or made available for his use. Nevertheless, the contractor can make use of the advantageous position he is in, partly due to a feeling of commitment, and because he is the first to know about discoveries made during his consultation with the professor or researcher. The contractor is also in a favourable position with regard to access to undergraduates and MSc or PhD candidates.

Operation of the scholarship grant system Table 3.4 indicates the trends in the operation of the scholarship grant system in national universities over a recent 8-year period. The number of cases increased at an average annual rate of 14.3 per cent in the 8 years FY 1975 to FY 1983. On a monetary basis, the increase totalled 16.5 per cent. This is a substantial rate of increase when compared with the 13.2 per cent rate of increase in research disbursements by industry and the 11.6 per cent rate of increase in research disbursements for the country as a whole over the same period.

Table 3.4 Trends in the scholarship grant system, Japan

			Foreign		*Domestic*	
Year	*Cases*	*¥ millions*	*Cases*	*¥ millions*	*Cases*	*¥ millions*
1975	8,082	4,428	54	114	8,748	4,800
1976	10,218	4,642	48	78	10,175	4,564
1977	11,414	5,522	51	86	11,363	5,436
1978	12,821	6,595	38	60	12,783	6,535
1979	14,102	8,095	30	43	14,072	8,032
1980	15,972	10,165	23	60	15,949	10,105
1981	18,606	11,658	28	42	18,578	11,616
1982	21,137	13,168	17	33	21,120	13,135
1983	23,525	14,994	24	38	23,501	14,956

Source: Ministry of Education

The average size of the scholarship grants made in FY 1983 was ¥637,000. While this is not a very large sum, it must be borne in mind that instances of industry–academic contact under this system within national universities in this year numbered over 23,500, and that this is rising at the rate of roughly 14 per cent annually. It should also be noted that foreign corporations are also making use of this system, although on a less frequent basis.

According to the MOE, the operation of systems for co-operation between industry and national universities resulted in an inflow of R&D funds from industry to national universities totalling ¥15,646 million in FY 1983 (Table 3.5).

The bulk of the R&D funds from the government are used for personnel costs and basic operating expenses. As a result, university researchers need an outside source to shoulder the burden of expenses which are accrued in the execution of their activities. To meet these needs, subsidies for scientific research expenses have been provided for within the MOE budget.

Table 3.5 Salient statistics of industrial–academic interaction at the national universities, Japan

Type of interaction	*Cases*	*Value (¥ millions)*
Contract research	241	302
Co-operative research	56	350
Scholarship grant	23,525	14,994
Total	23,822	15,646

Source: Ministry of Education.

The ¥39.5 billion total for scientific research expense subsidies in FY 1983 was allocated to 13,577 different research themes. The minimum allocation was ¥100,000 and the maximum ¥85 million, with the average allocation amounting to ¥2,909,000. These scientific research subsidies are allocated not only to national universities, but also to public and private universities.

Private universities, while not subject to MOE stipulations, have nevertheless set up various similar systems for the receipt of specific grants. These reported an inflow of ¥5,083 million in research funds from the private sector in FY 1983, which contrasts with the ¥14,994 million in scholarship grants awarded to those in national universities.

The Japan Society for the Promotion of Science

The functions of the JSPS, which is a special status corporation attached to the MOE, can be described as follows, in accordance with the Japan Society for the Promotion of Science Act, Law 123, promulgated on August 1, 1967: postdoctoral research fellowship support, incentive research fellowship/special fellowship system for collaborative research, special overseas research fellowships, research fellowship system for inviting foreign researchers, international scientific co-operation through agreements (exchange of researchers and research councils with developing countries based on mutual agreements), promotion of industrial–academic interaction, publication of research results, support for patent activities at national universities, donation operations.

The major concern here is the promotion of industrial–academic interaction. The JSPS set up a General Research Liaison Council in November 1982 to provide comprehensive support for industrial–academic research co-operation. This council is composed of up to 20 members, with knowledge and experience in academic and industrial circles. It deliberates on three themes: the mode of co-operation and collaboration between industry and academia in the years ahead, the selection of research topics anticipated to assume increasing importance in the future, and the establishment of special R&D committees. As a result of such deliberations, two special committees have already been formed, namely R&D on the self and third-party recognition of organisms (May 1, 1983–March 31, 1986) and R&D on intelligent language processing (April 1, 1984–March 31, 1987). Both of the committees are composed of specialists from industry and academia, and they thus constitute a forum for dialogue and interaction between the two spheres, with government subsidy. If industry decides during the course of the term that there would be value in more protracted co-operation in the field, it can be called upon to sponsor its continuation.

There is also a voluntary group, the JSPS Associated Society, set up within the JSPS through the co-operation of industry and academia. Industry shoulders the expenses, and there are about 40 research committees concerned with co-operation between industry and academia. The committee meetings are attended by representatives of both sectors, and the activities include the airing of opinions and reviews on technical problems, release of the results of testing and research, and field trips to factories and laboratories, all in accordance with the themes of the committee in question. Such a gathering has been interpreted as a venue for industrial–academic interaction in the broad sense of the term. It is mainly a forum for discussion since it does not conduct any actual testing or research.

Note 1. It should also be noted that in addition to the above, various other opportunities for public and private interaction have been created, adopting the philosophy that it is beneficial for representatives of industry and academia to exchange opinions and provide each other with food for thought about specific themes.

Science and Technology Agency interaction policies

There are basically three industrial–academic interaction systems promoted by the Science and Technology Agency (STA). The first is through the Research Development Corporation of Japan (RDCJ), a special corporation. The second consists of research contracts and trial manufacture contracted by STA special corporations to private corporations such as the Power Reactor and Nuclear Fuel Development Corporation, the Atomic Power Research Institute, and the NASDA, for the promotion of advanced technological development on a national scale. The third is the interaction arising through systems of contract and collaborative research and research personnel seconded from private corporations to the STA national laboratories, and the Institute of Physical and Chemical Research, another special corporation of STA.

An account will be given of the second type in Chapter 4, and the third has already been described under MOE interaction policies. This section will, therefore, consist of an account of RDCJ industrial–academic policies.

The RDCJ contributes to industrial–academic interaction through the following three operations: contracting of new technology development, new technology mediation, and promotion of creative scientific technology.

RDCJ system of development contracts

Outline The RDCJ was set up in 1961 as a special corporation under the jurisdiction of the STA. As of 1984 it had supported development on 259 different themes through the unearthing of research results obtained by a wide range of organizations, such as universities, national and public laboratories, private laboratories, and private corporations, with a view to leading these results to commercialization (Table 3.6).

The RDCJ selects as its themes those research results that entail great developmental risks and are difficult to commercialize, and contracts these to the private corporations responding to its

Table 3.6 Distribution of research in relation to Japanese organizations

Organization	*Cases*	*%*
University	151	57
National and public laboratories	64	24
Private laboratories	30	11
Independent persons	11	4
Industry	9	4
Total	265	100

Source: Research Development Corporation of Japan.

public offer. The organization provides the required facility and operating costs. If the development is successful, the corporation must repay this funding in annual instalments (interest-free) within five years. If unsuccessful, no repayment is necessary.

If the corporation succeeds in the development and wishes to proceed with commercialization, it concludes a use contract with RDCJ, and pays a fee to the RDCJ scaled to sales. The RDCJ in turn remits half of this to the owner of the new technology.

Operation of the system for development contracts The themes for development contracts are chosen not only from universities, but also from national and public laboratories, private laboratories, individual efforts, and even from industry. A selection is made of suitable private corporations desiring to implement developmental research. Most of the themes are chosen from universities (57 per cent of the total), followed in order by those from national and public laboratories (24 per cent) and private laboratories (11 per cent). This system functions effectively to promote industrial–academic interaction.

The rights of the university researchers and other new technology owners are guaranteed through the payment of user fees received via the RDCJ.

RDCJ system of new technology identification

Outline RDCJ issues invitations for selecting new technology possibilities to universities, laboratories, and others. The new technology is assembled and the prospective themes and topics of mediation aimed at commercialization selected. These are then introduced to corporations likely to have an interest in them.

The terms, such as the use fees and scale, are spelt out in a mediation contract concluded upon mutual discussion.

In the event that the corporation receiving the mediation requires funds for development, a portion can be borrowed from the RDCJ 'mediation fund'. After a deferral of two years, this must be repaid in annual instalments (interest free) within five years, and the corporation must also pay a special mediation fee equivalent to 20 per cent of the use fee.

Ordinarily, if technology gained through mediation is put to use, the special corporation claims 10 per cent of the use fee as a mediation fee, and transfers the remaining 90 per cent to the new technology owner.

Operation of the technology mediation system Since the inauguration of the mediation system between 1961 and 1985, there have been 235 cases of mediation (Table 3.7). The RDCJ currently mediates about 30 cases annually.

Table 3.7 Distribution of technology mediation system in relation to research organizations

Source of themes	*Cases*	*%*
University	49	21
National and public laboratories	105	70
Private laboratories	8	3
Independent persons	6	3
Industry	7	3
Total	235	100

Source: Research Development Corporation of Japan.

In this instance also, the new technology that is introduced through this system to private corporations for practical utilization is selected not only from universities, but from the national and public laboratories also. This system thus plays a major role in the overall system of industrial–academic interaction.

The Creative Science and Technology Promotion System

The Creative Science and Technology Promotion System involves the selection of talented individuals to stand as general project leaders for the unearthing of the seeds of scientific technology which may constitute the source of technological innovation. Under their direction, skilled researchers from the public, private, and academic sectors, as well as from abroad, are brought together to conduct probing research rich in creativity. It is the aim of this system to allow researchers full of creative instincts to exercise their talent to the maximum, with the focus on individuality. The name of the project leader is affixed to the project team.

Research themes range over a wide area, from material sciences to life sciences, with a research term of about five years. Seven projects have got under way since 1981, and the research funds involved came to ¥2,210 million in FY 1983.

The list of project leaders in Table 3.8 includes university professors and presidents, corporation heads, and chief researchers from laboratories. The project leader in the case of the Hayashi project is Dr Hayashi, president of the Japan Vacuum Technology Co. The leaders of the Masumoto, Ogata, Nishizawa, and Mizuno

Table 3.8 Projects under the Creative Science and Technology Promotion Programme

Project leader	*Title of project*	*Starting*
Hayashi	Ultra-fine Particle Project	1981
Masumoto	Amorphous and Intercalation Compounds Project	1981
Ogata	Fine Polymer Project	1981
Nishizawa	Perfect Crystal Project	1981
Mizuno	Bioholonics Project	1982
Hayashi	Bioinformation Transfer Project	1983
Horikoshi	Special Microbial Project	1984

Source: Research Development Corporation of Japan.

projects are university professors. The Horikoshi project is headed by Dr Horikoshi, a senior researcher (former university professor) of the Physical Science Research Institute.

The project leaders draw together the creative researchers needed to promote the research from private, public, and academic sectors. On average, the project teams are composed of four subgroups. The number of researchers assembled per project does not exceed 30, and the research plans envisage an expenditure of ¥2 billion to ¥3 billion in the five-year term.

The results are announced at meetings held under the sponsorship of the RDCJ each year, with applications for industrial rights made in some cases. The patents involved become the common property of the RDCJ and the individual inventor.

In the operation of the research, the project leader sets up a special office for the direction of the research with a staff to assist him in this task, and the RDCJ functions as the overall coordinator.

This system is currently rated highly by the government. Even though the FY 1985 national budget increased only 3.7 per cent compared with the previous year, and that portion related to science and technology only 3.0 per cent, the budget for this system amounted to ¥2,570 million, equivalent to an 11.9 per cent increase over the previous year, indicating the priority accorded it.

Universities have also warmly welcomed this project. There is sufficient monetary backing for the research, and since the nucleus is basic research, the work can be conducted from the standpoint of scientific interest. At the same time, there is an ever-present drive towards applicability in industrial technology in the interaction between the three sectors. This leads to an understanding of the extent of the application of the principles and phenomena discovered in the course of the work in the areas of engineering or industrial technology, which heightens intellectual curiosity even more.

Since Japan's industrial executives recognize the importance of technology development based on the fruits of unique and basic research, they have great interest in this system. In fact, representatives of industry also participate in some of the subgroups of the above projects. Some examples are as follows:

The Hayashi project consists of

(i) the Basic Material Properties Group (Nagoya, Meijo University),
(ii) the Physical Application Group (Ibaraki, Tsukuba Research Consortium),
(iii) Biochemical Application Group (Saitama, Maruzen Oil), and
(iv) Generative Methods Group (Chiba, Japan Vacuum Technology).

The university teams conduct the basic research, and the industrial teams target-oriented basic research and applied research.

The Masumoto project is composed of

(i) the Basic Material Properties Group (Sendia, Research Institute of Electromagnetic Alloys),
(ii) the Non-crystalline Compound Materials Group (Tokushima, Ozuka Chemical),
(iii) the Non-crystalline Film Materials Group (Tokyo, Gakushuin University),
(iv) the Special Ceramics Materials Group (Tokyo, Furukawa Denko), and
(v) the Intercalation Compound Materials Group (Sendai, Research Institute of Electromagnetic Alloys).

The other projects also involve participation by both industry and academia, with the research carried on at different locations at the subgroup level, but with the total effort greatly stimulated through the exchange of information and views within the project.

The active participation of industry is presumably spurred by a desire to apply the results over the long term.

The Research Interaction Promotion Act

As already noted, the Science and Technology Sub-Committee of the Council for the Promotion of Administrative Reform, in its report to the Prime Minister concerning future scientific and technological policy submitted in July 1985, advised on a legal system to promote research interaction. In response, the STA set up a council to deliberate on such measures. This council is to draw up legislation for the promotion of both industrial–academic and international interaction, to be put before the Diet. The substance and effects of this proposed legislation are thus topics of the future.

MITI interaction policies

MITI operates with basic researchers as part of its industrial technology policies. While these policies involve interaction between industry and academia, the emphasis is on national laboratories rather than universities.

These interaction policies are conducted mainly through the AIST and the 16 national laboratories attached to it and the Japan Industrial Technology Association (JITA). MITI operates the following systems: Technical Guidance and Consultation; Japan Industrial Technology Association (JITA), involving (a) Technology Transfer and (b) the Major Research Topics Study Committee; the Next-Generation Industrial Technology Promotion System; and the Basic Technological Research Facilitation System. Additionally, a parallel system is operated by MITI with the Ministry of Education for research contracts, collaborative research, and secondments from industry.

Technical Guidance and Consultation

The Technical Guidance and Consultation System was set up in 1957 to benefit those corporations that seek to exploit the technology and know-how accumulated over many years by the 16 national laboratories under the jurisdiction of AIST, and which desire technical guidance or consultation. Technical consultation is extended free of charge, but there is a fee for technical guidance. In 1983, 938 cases of guidance and 12,080 cases of consultation were undertaken, showing the frequent contact made by industry. The use of these systems has been increasing over a five-year period. Cases of guidance and consultation totalled 581 and 9,520 respectively, in 1979, indicating a 1.6- and 1.27-fold increase over five years.

Industry is attracted by these systems because of their easy availability. In FY 1983 the AIST system of contract research was used only once, whereas that of collaborative research was used a total of 18, 37, and 57 times in FY 1981, 1982, and 1983, respectively. These figures serve, by contrast, to illustrate the popularity of the technical guidance and consultation systems.

Public laboratories not attached to MITI but to local governments also conduct a lively technical guidance and consultation business, as described in Chapter 4.

Japan Industrial Technology Association

The role of the JITA consists of furnishing AIST R&D results to industry, investigative research for AIST R&D, dissemination of information about AIST R&D activities and results, backing for technological interaction between the national laboratories and industry, and industry–government liaison committees and symposia, as well as other planning for mutual understanding between national laboratories and industry.

Technology Transfer Since 1969 JITA has accumulated 19,748 instances of AIST-related patents and technology, of which a total of 1,143 AIST patents and 1,348 technologies have been transferred to 347 companies. Most representative technologies transferred from JITA to private companies include:

- isomerase technology (Fermentation Research Institute),
- melibiase technology (Fermentation Research Institute),
- high-performance electricity conductive film employing indium oxide as the main ingredient (GIRIO),
- PAN carbon fibre (GIRIO),
- sea water desalination technology (large-scale project),
- electron-beam exposure equipment (Electro-technical Laboratory),
- magnetic valves (large-scale project),
- powder magnet manufacturing method for metallic compounds (Electrotechnical Laboratory), and
- Na–CMC manufacturing method (NCLI).

In the dissemination of AIST patents, JITA is active not only domestically, but also dispatches missions overseas for interaction with the most advanced fields of technology, gives presentations about this technology, and participates in international exhibitions of technological exchange.

Up to the present time, the use ratio of the patents under its jurisdiction is 5 per cent.

Recent cases of big technology transfer by JITA reportedly include pitch carbon fibre (Government Industrial Research Institute, Kyushu) and microbial generation of γ -linoleic acid (Fermentation Research Institute).

The following means are taken to promote such technology transfer: meetings to explain the patents and presentation of the technology; visits to the corporations by consultant engineers to promote technology transfer; upon conclusion of an option contract, detailed technological information is passed from the AIST laboratory and the company commences R&D; release of AIST research results through symposia, publicity meetings, and seminars; specific reports of AIST labs furnished upon request by the company; the *JITA News*; dispatching of technology transfer missions overseas, and overseas presentations of latest technology developed at AIST laboratories.

Major Research Topics Study Committee This committee studies the trend of the latest technology and makes a multifaceted examination of the direction of technological research in such areas as energy and new materials (fine ceramics, new high polymer, high-performance metals, industrial materials theory and development — 14 themes), pattern information processing systems, function elements, biotechnology (six themes), as well as the promotion of local industry and technology. The committee has already conducted studies of more than 50 topics. Through its work the government determines the orientation of themes for large-scale projects at AIST. The study committee is operated with the participation of industry, academia, and the national laboratories.

There are currently two organizations functioning as vehicles for industrial–academic interaction within these areas. The first is the High-Polymer Research Promotion Council, made up of 62 corporations contributing 90 members, and 11 AIST labs contributing 57 members. The second is the Bionics Research Promotion Council. Its membership is composed of 66 individuals from 37 companies, 92 from nine AIST labs, and 32 from 16 universities. It is the aim of these groups to search out the direction of future research through discussions about new technology anticipated to blossom in the next century.

System for the promotion of next-generation technology

This system was not in itself designed as a system of industrial–academic interaction. It was set up in 1981 as an extensive reinforcement of R&D efforts in the area of basic technology indispensible to the establishment of next-generation industry, such as aerospace technology, information processing, new energy development, and bioindustry.

The innovative, basic technology that is the subject of this system is of a 'breakthrough' nature, with far-reaching effects, aimed at upgrading the level of existing industry, and alleviating resource and energy restrictions. At the present time the system's themes fall into three fields: new materials, biotechnology, and new functional elements. These can be broken down as follows:

The Fine Ceramics Research Association
  High-performance ceramics
The Polymer Research Association
  Synthetic membranes for new separation technology
  Electricity conducting polymers
  High-performance crystalline polymers
The Next Generation Metal Composite Materials Research Association
  Advanced alloys with controlled crystalline structures
  Advanced composite materials
The Biotechnology Research Association
  Bioreactors
  Large-scale cell cultivation
  Recombinant DNA
The New Functional Devices Research Association
  Superlattice devices
  Three-dimensional integrated circuits (ICs)
  Fortified ICs for extreme conditions

The notable point of this system from the standpoint of industrial–academic interaction is the process of implementation, which involves the joint efforts of industry, academia, and the national laboratories. Although this is also the case with regard to other AIST developmental projects, the effective co-operation of industry and academia is especially in evidence in these next-generation projects.

The pattern of this co-operation is not always the same. But, in general, the AIST functions as the theme promotion office, and contracts the research to the association formed by the corporation through the use of national funds. Sometimes that portion of the research which is basic is further contracted to universities by the R&D associations. In addition, university professors also participate in committees formed to evaluate and manage the research. The national laboratories also receive funds earmarked for the themes in question from the national budget, participate in the project, and are allocated part of the research. Figure 3.1 indicates details of the biotechnology project as an example of the participation structure.

The project members for the three themes include the Biotechnology Developmental Research Association (composed of 14

Figure 3.1 The Biotechnology Project within AIST

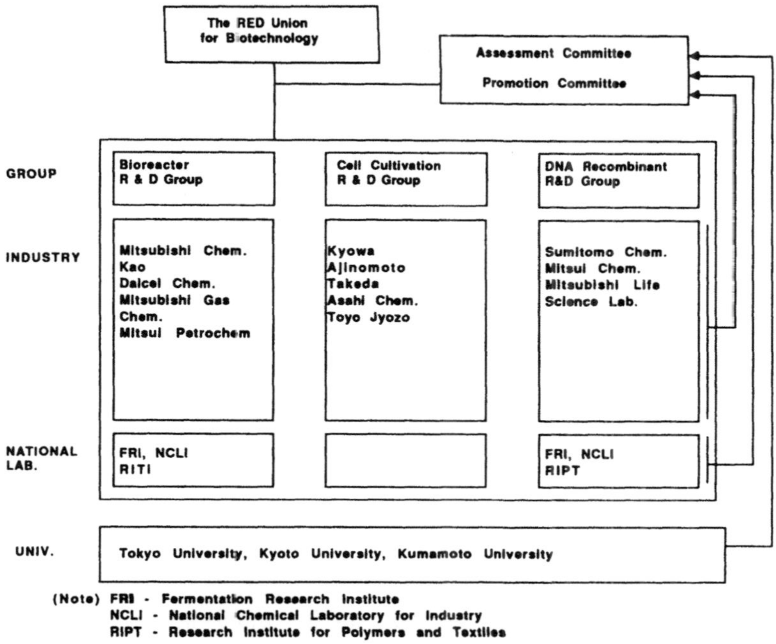

companies) and AIST laboratories (Fermentation Research Institute, Research Institute for Polymers and Textiles, and National Chemical Laboratory for Industry), with various sub-themes allotted among the members.

The basic technological research common to the three themes is contracted to an association of Tokyo University (Pharmacology Department, Agriculture Department, Applied Microbiological Research Institute), Kyoto University (Engineering Department), and Kumamoto University (Department of Medicine).

As noted above, university professors sit on evaluation and research management committees. It should also be noted that a special budget is approved within the contract under the heading of 'general study expenses', for discussion meetings with university professors on a per-group basis, roughly 10 times a year.

This system functions as a body for research contracts officially funded entirely by the government. Consequently, the research results are dealt with in accordance with the provisions of National Property Act, meaning that patents arising through the consignment revert to the government.

The composite materials R&D project can be cited as another example. According to Mr Kida, a division head of the Next-Generation Metal Composite Material R&D Association, those participating in this project number 236 in total, 143 from industry, 60 from the national laboratories, and 33 from academia (see Table 3.9).

Table 3.9 Participants in the Composite Materials R&D Project

	Industry	*National laboratory*	*University*	*Total*
Consigned[a]	31	19	7	57
Laboratory work[b]	101	30	3	134
Total	132	49	10	191
Consigned survey for next-generation technology	11	11	23	45
Grand total	143	60	33	236

Source: Next Generation Metal Composite Material R&D Association.

[a] Overall research co-ordination

[b] Laboratory research.

Although the two examples differ in scale, both provide a vehicle for the interaction of various individuals around the project, hopefully stimulating creative research. Therefore, although the actual research efforts are carried out in different locations, the projects have built-in joint industrial–academic meetings to study certain research themes and exchange information.

System for the Promotion of the Technology Infrastructure

This system was set up through legislation in 1985, and is promoted by the MITI. The Basic Technology Research Promotion Centre is set up as the arm of this promotion, and the system initiated the following operations beginning in 1985: financing (conditional interest-free loans); investment; promotion of collaborative research; testing and research through a concerted effort of industry, academia, and government; invitations to foreign researchers (Japan Trust); supply of research information; investigations. The aim of the system is the promotion of basic technology. In practice it will propose collaboration between industry and universities, issue invitations to foreign researchers, and monitor the supply of research information.

Since the start of the 1980s the keynote of scientific and technological policy in the advanced countries has been a stress on basic technology, and Japan is no exception. In this vein, Japan has hammered out successive policies for the promotion of basic technology, as evidenced by the Creative Science and Technology Promotion Policy in 1981 (Chapter 3, p. 70) and the Next-Generation Industrial Technology System (Chapter 3, p. 75).

While the MITI cannot extend its policy to the national universities under the jurisdiction of the MOE, it is considering a more effective usage of the 16 national laboratories under the control of the AIST, which also come under the jurisdiction of the MITI. Along with funds from the private sector, this system, through the collaboration of the MITI and the Ministry of Posts and Telecommunications, will acquire revenue from dividends of NTT (the counterpart of British Telecom), the status of which was changed from special corporation to joint stock corporation. It diverts these to basic research funds. The framework of its operations is planned as follows:

Financing and investment Supply of risk money needed for technology development in the private sector. Not a system of industrial–academic interaction.

Promotion of collaborative research Promotion of collaborative research between national laboratories and industry, particularly in the advanced or interdisciplinary fields of technology, through: supply of research-related information in industry and academia needed for planning of collaborative research projects, proposal of collaborative research projects, and execution of various procedures needed for collaborative research for the parties involved.

Testing and research involving the concerted efforts of industry and academia The Basic Technology Research Promotion Centre accepts contracts from private corporations on themes in which the latter has needs for technology development and the funds to do so, but is unable to commence the testing or research due to a lack of capabilities. The centre brings together researchers from the national laboratories and private corporations to conduct the research.

The themes in this case should concern basic technology, and the research potential of the national laboratories is available for the execution of the contract.

Invitations to foreign researchers Selection of researchers to be invited, co-ordination with the organ to receive them, extension of financial aid and subsidies for the invited researchers, and the formulation of invitation plans. As a general rule, the reseachers are invited to pursue their work at a private corporation for a year.

Investigations and supply of research information Investigations to promote testing and research activities related to basic research and the supply of such information to the private sector.

This System for the Promotion of the Technology Infrastructure is not yet in full operation. Its effectiveness can be said to hinge on the efforts made by those concerned in the years ahead. The funding available to it appears to total just over ¥4 billion for a year, but is certain to increase.

Local government interaction policy

The 47 prefectures (equivalent to British counties) and 3,255 municipal and other local bodies discharge various government functions at the local level. Each of the prefectures has public laboratories for the promotion of local industry, and testing and research for the common good. There are also such laboratories in the big metropolises.

To serve the purposes of industrial expansion at the local level, these public laboratories conduct the following services: autonomous testing, research, and investigation; contract testing and research; courses, lectures, provision of speakers, technical guidance, and other similar operations; facilities; and technical consultation. Users of these services are not only small and medium-sized but also big corporations.

Here an account will be given of the Osaka Municipal Technical Research Institute as an example of these public laboratories.

Osaka Municipal Technical Research Institute

The OMTRI has a history of about 70 years. It is staffed by 95 researchers, 17 managers, and 11 apprentices, a total of 123. Its research work concerns organic and inorganic chemistry, biochemistry, plastics, and mechanics. Its operational indices are shown in Table 3.10. Some notable features are its great volume of contract research, testing, and analysis, and the technical consultation it conducts in a laboratory staffed by 95 researchers.

The research personnel of OMTRI publish about 50 reports annually. General surveys and other writings number about 20 annually, as do inventions. Inventions and patents in most cases result from collaboration with private corporations, and contracts

Table 3.10 The activity of OMTRI

Type of activity	*Number of cases*
Self-supporting research	35
Consigned research[a]	
A	186
B	100 (105)
Total	286
Testing and analysis (for fee)[b]	3,103
Technical consultation (no fee)[c]	13,457
Renting equipment for test and laboratory work	119
Use of library	1,355

Source: Osaka Municipal Research Institute Association.

[a] No researcher is dispatched from the consigner in the case of A, while one or two researchers are dispatched in the case of B. The number of researchers dispatched from the consigners is in parentheses.

[b] Testing and analysis of sample materials are provided by private companies.

[c] OMTRI researchers discuss or give advice on specific technical issues disclosed by private companies.

regarding their use have been concluded with the companies in question.

One of OMTRI's roles in industrial–academic interaction is the guidance and the dissemination of technology. It holds institute classes, conducts an itinerant technical guidance service, and trains engineers in industrial technology. It also accepts technicians from developing countries for technical training (courses in plastics and fermentation engineering) on contracts received from the Japan International Co-operation agency (JICA).

One of the affiliated groups connected with OMTRI is the Osaka Municipal Technical Research Institute Association (OMTRIA). This association has about 450 corporate members and holds courses, lectures, and exhibitions. Its monthly magazine is titled *Kagaku To Kogyo (Osaka)* — literally, *Science and Industry*.

These functions and the membership subscriptions to *Science and Industry* provide a forum for interaction and mutual enlightenment. OMTRI researchers participate actively and contribute articles to the magazine. The industrial–academic interaction around OMTRI can be held up as an example of a successful operation.

In general, the other local public laboratories can be regarded as having operations similar to those of OMTRI, though with varying levels of activity.

Since it is a municipal organization, some 75 to 80 per cent of its partners are corporations within the municipality. Nevertheless, 6 to 11 of the corporations come from outside the municipality to avail themselves of OMTRI's services (Table 3.11).

Table 3.11 Locations of companies associated with OMTRI

Type of activity	*(1)*	*(2)*	*(3)*
Contract research	80	13	6
Testing	78	11	11
Technical consultation	74	15	11

Figures are given in percentages.
(1): Companies located in Osaka City.
(2): Companies located in Osaka Prefecture.
(3): Companies located in prefectures other than Osaka.

Osaka City disburses approximately ¥1.1 billion yearly to support OMTRI for the promotion of local industry. Although consigned R&D and other systems operate on a fee basis, revenue gained in this fashion is not sufficient to operate the institute. OMTRI is supported by fees from member corporations. Its FY 1983 budget was ¥62 million.

Britain

General initiatives

Steps have been taken in recent years to increase the interaction between institutions of higher education and industry: 'research parks' have been created; industrial liaison officers have been appointed in universities; consultancy companies have been established; science parks and innovation centres have been set up; conferences, seminars, and exhibitions have been held to promote collaboration. Chapter 6 (p. 180) provides a full list of such developments in Britain to September 1986. Third parties such as the Science and Engineering Research Council (SERC), the Department of Trade and Industry (DTI), the North Atlantic Treaty Organization (NATO), the European Community (EC), and the British Technology Group have also been active in improving the flow of technology transfer.

Research parks

The main difference between a science park and a research park is that in the terms and conditions of tenancy of the latter, only research and not production may be carried out. As many companies prefer to conduct research alongside their production facilities, science parks have become more numerous than research parks. Heriot-Watt Research Park is a good example of an effective centre that has maintained its emphasis on research, and fully utilized the resources of the university nearby.

Industrial liaison officers

Apart from research, universities and colleges are able to offer to industry other services such as technical and business information, from the development of processes and products to the actual marketing of innovations. Industrial liaison officers are wholly sponsored by the college to provide 'services to industry' for the benefit of the college. Functions may include: facilitating industry access to college facilities (e.g. testing facilities, special equipment, staff as consultant, etc.); advertising and/or selling courses to industry; aiding student employment during sandwich courses.

Consultancy company

A natural extension of the activity of industrial liaison officers is to form a limited company to provide services to industry. A good example of such a company is Salford University Industrial Centre Limited. The centre's objective is to offer an efficient, cost-effective, and confidential service to provide clients with the right technological package, from solving a specific scientific problem to developing the germ of an idea into a successful product. The service draws on the facilities and knowledge of the university in the key areas of electronics, engineering design, manufacturing technology, energy conservation, noise control, and chemistry.

Science parks and technology parks

A further step down the road of interaction is to utilize spare land and buildings to provide space for technologically based firms to start up and grow, in a campus-like setting close to a major college, and though they are mainly associated with a particular university, they tend to have a national or international orientation.

The Cambridge Science Park is the oldest and most successful of all the science parks in Britain. In recent years many universities throughout Britain have tried to emulate the success of the Cambridge Science Park. Some have been more successful than others, though none have come anywhere near the success of Cambridge. (A review of such science parks and particularly their role in promoting high-technology firms has been undertaken by Peak, Marwick, and McLintock on behalf of the UK Science Park Association; the results are detailed in Chapter 8.)

There has been a tendency in Britain for universities to concentrate on the land and buildings, and correspondingly insufficient attention has been devoted to the technology transfer aspect, that is the means of identifying and then mobilizing a university's resources and developing a particular project into a commercial venture.

Business and innovation centres

The building of 'innovation centres' with adequate support staff and links to industry and academic departments could prove a more practical form of technology transfer than the large science parks. This development has recently been vigorously pursued in Europe, where a network has been formed to share experience and new developments. Since this movement does not appear to have a counterpart development in Japan, more detail will be presented later about one such centre (Newtech Ltd; Chapter 6).

Technology Clubs

Several TCs have been established in various parts of Britain. One of the most recent is the Cardiff Technology Club. The club, initiated by the University of Wales Institute of Science and Technology, is designed to provide a forum where industrialists and researchers can meet on a regular basis to discuss problems of technology interests. It is hoped that this will stimulate a more rapid transfer of high technology from university centres to industry. Apart from the discussion, each meeting will be addressed by an eminent speaker drawn from industry, a university, or government agency sources.

Conferences

Many academic establishments throughout Britain hold conferences with a view to transferring technology. The Research Division of the North East Wales Institute of Higher Education can be taken as one example. Since 1974, the Institute has played host, either at its Wrexham or Deeside campus, to local, national, and international meetings. The topics covered in these symposia have been diverse and have included skin photosensitization by drugs, developments in radiation chemistry and biology, gums and stabilizers for the food industry, electron-organic synthesis, and the Cellucon series of conferences. Cellucon-88 was held in Japan, jointly with the Society of Fiber Science and Technology of Japan. These have been most successful. In particular, the meetings have served as useful interfaces between industrialists and research scientists. The research division of NEWI, therefore, has gained substantial expertise in symposia organization, and is able to cover all aspects including advertising, circularizing, scientific programme, social programme, catering, accommodation, press and television releases, financial administration, and so on. This service is aided by an efficient computerized system that deals exclusively with all these organizational aspects.

Third-party initiatives

Science and Engineering Research Council

The primary function of the SERC is to maintain and enhance the research capability of British universities in all areas of pure and applied science and engineering, outside those covered by the more specialized research councils. The SERC's basic research responsibilities consequently extend over a very wide range, from the most fundamental to the severely practical. The council has pioneered ways of bringing the universities and industry closer together in research on engineering and applied problems of national importance.

In allocating its resources, the SERC looks to the peer review system to achieve and maintain high standards of research and to support programmes regarded as timely and promising. The SERC and its boards and committees are made up of representatives from institutions of higher education, industry, and government. Membership is subject to regular review.

By identifying areas of practical importance and co-ordinating the efforts of academic and industrial researchers in an active way, both the intellectual content and the practical effect of university research have been greatly improved. 'Focused' research of this kind has been carried out within the council's specially promoted programmes and directorates, while more recently the Science Board has identified 'themes' within core science for special support.

In addition, many activities that the council supports are co-ordinated with those of other major spending organizations whose interests are complementary with those of the council. Joint research programmes are undertaken, in particular with government departments and nationalized industries.

The six SERC schemes directly aimed at supporting education–industry interaction are the following.

Co-operative grants The SERC introduced its Co-operative Grants Scheme in 1978 in order to encourage direct industrial involvement in academic research work at an early stage. Under the terms of the grants, the SERC's contribution may be up to three times that of the industrial partner in terms of direct costs. A firm that contributes financially, or through effort or equipment, at least half the resources needed by an academic project receives the rights in the development work. The scheme has some 180 projects already in progress, to which the SERC contributes about £1.5 million p.a., and a wide variety of firms are involved.

One noteworthy grant of £182,653 was to Leicester University to support the formation of the Leicester Biocentre. The co-operating partners in this grant are four industrial companies: John Brown Engineers, Gallahers, Spillers-Dalgety, and Whitbread, which together will contribute £1 million over five years to the university to establish the Biocentre. The Wolfson Foundation has agreed to provide £570,000 for a building to house the centre, and the UGC will provide £175,000 to meet fees and furnishings for the building. This is a unique academic/industrial/governmental collaborative venture to support an initial research programme in the molecular genetics of yeast and plant cell culture and plant biochemistry.

A similar programme in biotechnology at the North East Wales Institute has led to the establishment of a new company, Biokits Ltd.

Co-operative Awards in Science and Engineering The CASE scheme offers a type of research studentship where an industrial organization joins with the university or college in defining a project and supervising the students' progress. The company also agrees to make its own facilities, including appropriate laboratory training, design, and manufacture, available to the student for a part of his or her total training period.

This is a method of giving the students a new perspective on the significance of their research in an industrial context, with a project selected jointly by the academic partner and industry. It leads to a type of training directly relevant to the industry's interests and training for its staff. In addition, the resulting contact between academic and industrial supervisors can generate new interaction. Seven hundred and sixty new CASE awards were made by the SERC in 1983, representing about one-third of all its Ph.D. awards.

Collaborative Training Awards In 1979 the SERC introduced the Collaborative Training Awards Scheme to support short-term training projects of 12 to 15 months which were more related to the problems of design and production and more relevant to the needs of smaller companies. The number of projects launched, however, has been small.

The Integrated Graduate Development Scheme The Integrated Graduate Development Scheme provides financial support to new graduate recruits to the manufacturing industry for intensive induction training, organized jointly by the participating companies and a designated university or college. The training is provided during alternating periods at the companies and the academic institution, blending tuition with practical experience at the factory level.

The Teaching Company Scheme The Teaching Company Scheme began in 1975 and is jointly funded by the Department of Industry and the SERC, and has proved to be very successful in promoting technology transfer.

It aims to encourage well-qualified first-degree students to develop production engineering skills (to a M.Sc./Ph.D. standard) by working in a company for a 2- to 3-year period, while also improving the productive efficiency of the company. The student (or associate, as they are called) is company based, with an academic supervisor from college who can arrange trainee tutoring.

The Royal Society/SERC Industrial Fellowship Scheme The Royal Society/SERC Industrial Fellowship Scheme was launched in 1981 to enable academics seconded to an industrial environment to undertake a project, lasting from six months to two years and related to the company's R&D programme, and to enable industrialists to undertake research or course development in a university or polytechnic. The fundamental aim is to improve understanding of the problems of conducting R&D in the other environment. In the first 18 months only 21 applications were received and 15 awards made, but it has been a general experience that all such schemes take time to gather momentum.

British Technology Group

The BTG's objective is to promote the development of new technology in commercial products, particularly where the technology originates from public sector sources such as colleges, research councils, and government research establishments.

The BTG corporate plan has now been approved by ministers. Under this plan, the BTG will concentrate its future efforts on technology transfer, helping industry to exploit technology from British public sector sources. More specifically, the BTG will: offer to take responsibility for patenting, or otherwise protecting, technology derived from universities and other public sector sources; provide funding for the development of the technology to the point where it can be taken up by industry; transfer the technology to industry by seeking licensees; offer project finance to help those licensees to launch the product onto the market; share the licensee income with the source of the technology (either the institution or the inventor); plough back the retained share of the licence income into the development and exploitation of other technology.

As part of its technology transfer role, the BTG will also offer project finance to companies that want to develop new products and processes, based on their own technology.

In the cases where a particular technology requires the setting up of a new company, the BTG will perform a catalytic role in promoting the creation of start-up companies.

All these activities will be carried out on commercial terms. The BTG expects to be both profit-making and self-financing throughout the period of the five-year plan (1984–89).

The corporate plan expects the BTG to invest about £15 million per year from its own resources. This will enable it at least to double its current rate of investment in development projects at universities and other public sector sources.

In addition to its traditional business of supporting the exploitation of inventions through licensing, the BTG will in future provide finance for other methods of technology transfer. The BTG will offer up to 50 per cent of the finance required for projects to support contract R&D, consultancy work, and other services provided by university departments and industrial liaison companies. The finance will be offered on commercial terms, and the BTG will expect to recover its investment from the revenues generated by the institutions concerned. A new initiative called 'Campus Investments' will specialize in providing finance for academic spin-off and start-up companies.

Inventors and potential licensees will benefit from a new computer database that the BTG is compiling to assist the preliminary matching of new ideas with possible industrial users. In future the BTG will share royalty income with all sources of inventions. It will also increase the share of royalty income paid to the source during the early stages of the exploitation of an invention. Forty-two universities have agreed in principle that in future the leaders of BTG-funded projects will be designated as BTG Research Fellows.

The BTG has set up an eight-man liaison team (full-time and part-time) whose role will be to seek out commercial opportunities by visiting laboratories throughout Britain. The BTG will also introduce regular portfolio reviews in order to improve communications and keep institutions better informed of progress in exploiting their inventions.

Since the establishment of the BTG (formerly NRDC), over 23,000 inventions have been offered from public sources, of which about 8,000 have been accepted for exploitation. Over 1,000 have been revenue-earning; about 30 have each earned more than £100,000, several over £1 million, and one over £100 million.

Finally, the BTG runs an annual Academic Enterprise Competition, with total prize money of £100,000. Prize money will be given to the most promising academic researcher who has set up or intends to set up a new business. Prizes will also be given for other methods of transferring technology from academic institutions to industry.

Department of Trade and Industry

The department has five research requirement boards, each dealing with broad sectors of industry or technology. Each board is chaired by a senior industrialist and the membership is drawn from the department, industry, research associations, and universities.

The boards determine the objectives and balance of the DTI research programme. Drawing on the knowledge and experience of their members, the boards help to identify priority fields for action and support to meet industry's needs.

Each requirement board is supplemented by subsidiary committees. One of the most important committees for the purposes of this book is the Technology Transfer Services Advisory Committee. This committee has members drawn from industry, research organizations, trades unions, and universities as well as from the institutions that run advisory or consultancy measures on behalf of the government. The committee advises the department on the overall balance of these services and on new possibilities for transfer of best practice techniques and emerging academic knowledge across industry.

Apart from the boards and committees the science and technology component of the DTI is organized into the following organizations or groups (Table 3.12):

Chief Engineers and Scientists Group
Laboratory of the Government Chemist
National Engineering Laboratory
National Physical Laboratory
Warren Spring Laboratory
Air (Civil Aircraft Technology)
The Alvey Programme
Chemicals, Textiles, Paper and other industries
Electronics

Table 3.12 Science and technology expenditure by organization, 1984–85 (£000 net)

Type of expenditure	*Industry*	*Research associations*	*Universities*	*Natural Environment Research Council*	*United Kingdom Atomic Energy Authority*	*Other government departments*	*Other*	*Total*
Chief Engineer and Scientist's Group								
Advisory	40	0	210	0	0	0	31,250[a]	31,500
Technology transfer	19,360	480	400	0	990	0	770	22,000
Laboratory of the Government Chemist	3,360	0	70	0	0	860	10[b]	4,300
National Engineering Laboratory	30	70	500	0	70	0	12,680[b]	13,350
National Physical Laboratory	190	0	340	0	0	1,540	20,900[b]	22,970
Warren Spring Laboratory	0	0	140	0	0	0	3,120[b]	3,260
Air	15,960	180	1,160	0	30	1,514	0	32,470
Alvey programme	3,150	0	0	0	0	2,400	900	6,450
Chemicals, textiles, paper, and other industries	5,150	5,890	250	0	0	1,910	20	13,220
Electronics	39,140	350	20	0	630	4,030	30	44,200
Information technology	41,450	2,850	530	0	130	1,530	1,190	47,680
Maritime technology	3,600	30	110	0	150	0	6,580	10,470
Mechanical and electrical engineering	31,910	3,260	1,110	0	2,850	300	320	39,750
Minerals and metals	2,620	3,410	110	1,230	50	0	1,100	8,520
Space technology	4,310	60	0	0	0	4,610	55,160[c]	64,140
Vehicles	11,040	680	0	0	150	0	200	12,070
Total	181,310	17,260	4,950	1,230	5,050	32,320	134,230	376,350

Source: Department of Trade and Industry.

[a] Industrial consultants

[b] Intramural expenditure on DTI programmes.

[c] European Space Agency.

Information Technology
Maritime Technology
Mechanical and Electrical Engineering
Minerals and Metals
Space
Vehicles

The DTI's Chief Engineer and Scientist advises the ministers on science and technology matters generally. He is responsible for the department's support for industrial R&D and the management of the science and technology budget.

Two divisions of the department, Research and Technology Policy (RTP) and Quality and Education (QE), help the Chief Engineer and Scientist carry out his responsibilities and directly administer a number of measures of assistance for R&D, innovation, and technology transfer.

The RTP Division's responsibilities include: the development of the department's current and long-term science and technology policy; co-ordination of that policy with the science and technology policy of other departments; liaison with universities, research councils, and public sector organizations; policy towards European Community science and technology; policy co-ordination and central administration of the Support for Innovation scheme. The QE Division's responsibilities include: ensuring that industry has access to and is encouraged to exploit marketable research ideas generated in institutes of higher education; the provision of advisory and consultancy services to encourage firms towards improved profitability and a better climate for technological change; the department's interest in education and training to meet the needs of industry. Some of the more specific activities include the following.

Manufacturing consultancy Assisted consultancy is available over the whole range of manufacturing methods and organization. The management of this service is contracted to Salford University Industrial Centre (SUIC) Ltd for the northwestern region of Britain and to the Production Engineering Research Association (PERA) for the remainder of the country. Several hundred consultancy organizations provide the assistance required. More than 7,000 firms have been assisted since 1977, with more than 1,100 in 1984–85. The benefits have been shown to be substantial. As an example, a petrol pump manufacturer was helped to reduce work in progress by £250,000 increase its output by 25 per cent, and double its profits.

Quality assurance As part of its quality campaign the department provides assisted consultancy to small and medium-sized firms to encourage them to obtain certification of their quality assurance procedures to a nationally recognized standard such as BS5750. The management of the services is contracted to SUIC Ltd and PERA.

Six hundred firms were assisted in 1984–85. The assistance frequently produces some immediate benefit to the firm. For example, Stelrad, which makes domestic heating boilers, was helped to halve its rework costs and to reduce the number of products returned under warranty by 15 per cent. The firm intends to apply for quality assurance certification under BS5750 and eventually expects to reduce the cost of quality assurance by £1 million annually.

Design To encourage greater industrial awareness of the vital role that good design plays in profits and market share, the department assists small and medium-sized firms to make use of specialist design consultants. The scheme is managed on behalf of the department by the Design Council. Several hundred specialist design consultants provide assistance over the whole range of design, including engineering and the industrial design of capital goods, machinery, consumer goods, textiles, clothing, and food products. More than 2,000 firms have been helped in this way (1,000 in 1984–85), and many report a substantial increase in turnover as a result.

The most significant step taken in 1984–85 was the publication of the government's first-ever statement of design policy. This document, entitled 'Design: a Framework for Action', was launched in May 1984 and has led to the identification of priority issues. A strategy group has prepared a report responding to the government's initiative, and detailed proposals for action have now been prepared. At the same time there has been a continuing series of initiatives supported by the department including, alongside industry, the Royal Society of Arts' Young Designer into Industry scheme, under which newly graduated designers will have up to one year's programmed experience in industry to complete their training effectively.

Assistance to the Design Council, which is the government's principal agent for design, has continued. A grant of £4.2 million was made to the council to enable it to maintain its range of activities in the areas of service to industry-education, annual awards, production selection, marketing services, exhibitions, and publications.

Teaching Company Scheme This scheme establishes programmes between higher education institutions and companies, to bring about change in industry through the effective use of academic resources and advanced technology. High-quality graduates, called associates, are recruited to the companies for periods of two years to implement the change. The programmes also give academic staff direct involvement with industry, to the benefit of both their teaching and research. At the end of 1984–85, the number of programmes had increased over the previous year by 20 per cent to 165.

Funding for the scheme is provided by the department, the Science and Engineering Research Council, and industry. In 1984–85 the department contributed £2.3 million towards the total cost of about £7 million. Teaching company programmes work well in both large and small companies.

For example, Baker Perkins of Peterborough designs and manufactures a range of complex equipment for the food-processing and printing industries. They have become well known for their application of computer-aided engineering techniques and have operated a TCS programme with Loughborough University since late 1982. Within the overall manufacturing plan six associates have been employed on defined projects in areas such as planning, production control, tool management, and manufacturing cell management. Working relationships between the company, the university, and the associates have been excellent. In one project area they studied the manufacture of major cylindrical components for printing presses, then reorganized to take full advantage of the avilable technology. This resulted in a reduction of cycle time from 16 to six weeks, with an estimated saving of £250,000 per day. Equally important, most of the associates who have completed their two-year appointments have gone on to take senior posts with Baker Perkins. Following this success the company plans further TCS programmes.

By contrast, Surgical Equipment Supplies Ltd (SES) and the Polytechnic of Wales have completed a very successful programme with just one associate. SES is an autonomous company within the Smiths Industries Plc group, employing 65 people in the design and manufacture of a range of sterilization equipment such as autoclaves. TCS support was used to implement new sterilization techniques through revised product design incorporating microprocessor control. Here again working relationships have been excellent. In addition to the technical completion of the programme, the first resulting product gained the Underwriter's Laboratories' approval, enabling it to be marketed throughout the

United States without restriction. An additional British and European model developed by the same team was launched in August 1984, and 700 units had been sold by March 1985. The same partners have now enthusiastically agreed to a second TCS programme, while, encouragingly, the original associate has set up his own company, Hartstone Ltd, to design and manufacture medical electronic equipment.

Databases Technical and business information is made more available through two mechanisms. In the commercial sector, Support for Innovation is provided for the development stage of innovative databases and similar activities with potential for success in the marketplace. Half of the available funds of £5 million had been allocated by April 1985; 22 projects were supported in 1984– 85 with overall costs of £12 million; topics range from company financial data to techno-commercial data for the chemical industry. In the non-commercial sector, the department committed £330,000 to encourage information providers such as universities, research associations, and professional institutions to make their specialist knowledge available to industry through databases and other information activities. Project areas range from data on welding consumables or plastics materials to an information centre with on-line data access.

Education During 1984–85 the department had completed schemes to help primary and secondary schools purchase microcomputers and related equipment. Over 27,000 primary and 5,600 secondary schools have participated, and the number of micros in use in schools is increasing rapidly, with an average in secondary schools of at least 10. Teachers and children are rapidly developing skills in their use.

The department's Industry–Education Unit has helped educationalists to develop teaching resources, curricula, and examinations that relate to industry's interests in science and technology. This is an essential step in encouraging schools not only to stimulate the interest of young people in careers in engineering and industry, but also to help develop an awareness in the future population of the part that modern science and technology play in everyday life.

Reports on skill shortages in information technology were published by the department in July 1984 and January 1985. In March the government announced a programme involving expenditure of £43 million over three years for the provision of additional student places in higher education for engineering and science. Of this total, the department's contribution was £12.5 million.

Information Technology Centres The programme to create 175 Information Technology Centres (ITeCs) through the country is now complete, and the centres are proving of value in training 16–17-year-olds and some adults in information technology skills. At present there are no plans to increase the number of ITeCs which were set up with the aid of departmental pump-priming grant assistance of £75,000 per centre over three years. The department's involvement will continue until that period ends, and together with the Training Agency it is developing a marketing strategy for ITeCs' goods and services to enable them to generate revenue and create a stable financial base. The programme not only trains young people and some adults in IT skills, but also gives practical help to industry by creating with it an organization through which it can train and recruit staff.

European Community funded research programmes

The increased emphasis on European Community research programmes with industrial objectives, which started in 1984 with the establishment of the European Strategic Programme of Research and Development in Information Technology (ESPRIT), has continued. Of the projects announced in the first year of the programme, British organizations were involved in 70 of 109 projects; this included participation by 28 firms and 17 universities and polytechnics.

Two major new industrial programmes have been agreed. The programme on Basic Research in Industrial Technologies (BRITE) is designed to improve the technological base of and encourage the introduction of advanced technology into the more traditional industries in the Community. The programme supports collaborative R&D in the following areas: reliability, wear, and deterioration; laser technology and powder metallurgy; joining techniques; new testing methods; CAD/CAM and mathematical modelling; polymers, composites, and other new materials; membrane science and technology; catalysis and particle technology; and flexible materials. BRITE will cost the Community 125 Mecu (about £74 million), and although the successful schemes are not yet known, it is known that of the 3,037 potential applicants who expressed an interest initially, 861 were from Britain. Financial support from the Community will cover up to 50 per cent of the approved costs.

The second programme, on biotechnology, is designed to remove, by means of research and training, bottlenecks that prevent the exploitation of biotechnology in industry and agriculture. The programme will cost the Community 55 Mecu (about £33 million) over five years and covers measures to upgrade the quality and capacity of the infrastructure for biotechnology R&D; training and the pooling of facilities and knowledge in bioinformation and collection of biotic materials; enzyme and genetic engineering; technology of cells and tissue cultures in the laboratory; new methods for the assessment of risks associated with modern biotechnology.

The DTI has actively promoted the BRITE and biotechnology programmes by means of seminars and articles in order to encourage British industry and research organizations to participate fully in the European initiative. Further details of European-wide programmes are given in Chapter 7 (p. 189).

NATO — Scientific Affairs Division

In December 1982 the Scientific Affairs Division of NATO launched a new programme concisely called the Double Jump Programme. This programme is specifically aimed at promoting international collaboration between the researchers in universities and industrial laboratories. NATO is willing to sponsor projects of industrial interest.

In order to satisfy the specific requirements of the Double Jump Programme, co-operative research projects should (1) include scientists both from universities and industry and from different countries; (2) be of clear industrial interest which is witnessed by the willingness of the industry(s) concerned to provide the required financial support for its own scientists to carry on the project and to take part in the international collaboration whose academic portion is sponsored and financially covered by the NATO grant.

Government policy: the Green Paper on higher education

The Green Paper published by the Department of Education and Science in May 1985, 'The Development of Higher Education into the 1990s', goes much further than any previous government initiative to encourage higher education to meet the needs of industry more closely, and by so doing attract private-sector finance. These proposals have now been implemented.

The Green Paper paints a picture of higher education in the future in which universities are much more responsive to the needs of industry by (a) putting greater emphasis on science and technology at the expense of arts and social sciences; (b) providing courses that are more appropriate not only for undergraduates but also for mature students; (c) encouraging industry to take an increasingly important role in funding and guiding university activities; (d) by linking universities with local industry and commerce, not only to encourage the transfer of technology, but also to extend the use of artistic, cultural, and recreational facilities into the local community; (e) by fostering an entrepreneurial spirit 'essential for the maintenance and improvement of employment, prosperity, and public services'.

The Green Paper suggests that universities should encourage industrial liaison by allowing academics to take consultancy work; taking consultancy and other beneficial industrial work fully into account when assessing candidates for promotion; granting permission to engage in business activity on the basis of an assessment of its effect on an academic's overall contribution to the objectives of his institution; offering first-rate academics who have particularly strong and time-consuming links with business and part-time contracts or are seeking to make joint appointments with industry; striking reasonable deals with staff over the sharing of income earned from involvement with business; operating industrial liaison services; insuring against any liability for damage arising as a result of the advice offered by academic consultants to industry. In addition the government:

(i) has asked the UGC to ensure that universities' efforts and successes in industrial research and consultancy receive due recognition in the process of grant allocation and that universities are aware of this practice. This does not mean that universities' work for commercial clients should be subsidized, but that universities that are successful in carrying out such work should be helped to build on their strengths.

(ii) has confirmed that university earnings from collaboration with employers will not be penalized in the determination or allocation of UGC block grants. Earnings from the commercial exploitation of research results in all higher education institutions should be used to defray the costs of such work and to reinforce research.

(iii) has agreed to regard the costs of necessary infrastructure for collaboration with business as a proper use of an institution's general income.

(iv) has introduced legislation to remove barriers to commercial exploitation of their work by local authority institutions of further and higher education.

(v) has ended the British Technology Group's monopoly of first refusal on research council-funded inventions (this was announced on 14 May 1985). The government endorses the proposed transfer of responsibility for promoting commercial exploitation from the research councils to institutions of

higher education, with the delegation of this responsibility wherever possible to the individual researcher.
(vi) is working with the University Directors of Industrial Liaison (UDILs) and others to develop the UDILs' plans for a national data base for academic expertise and facilities in universities and polytechnics, as recommended in the ACARD Report; a commercial consortium led by Longman Cartermill Ltd (and backed by the Longman Group Ltd) has been selected to build and run the data base, to be known as 'British Expertise in Science and Technology'.

Chapter four

Industrial–academic interaction: assessment

Japan

The effectiveness of industrial-academic interaction can be assessed at the personal level and in terms of the financial benefits that accrue.

Human contact promotes an exchange of information. This exchange can be a one-way flow, but in many cases is two-way. The flow of information gives rise to discussion, which can result in the posing of problems and the emergence of ideas and hints for their solution. From this standpoint, the actual state of the human-centred industrial–academic interaction should be grasped by such barometers as the scale and frequency of the interaction. Although it was impossible to obtain overall indices, the results of questionnaires conducted recently provide an assessment of the situation. In summary, as shown in Table 4.1 the record of industrial–academic interaction for 534 Japanese corporations over five years indicates that:

(i) In the majority of cases, interaction with universities and national laboratories led to (a) guidance and advice for industrial R&D; (b) invitations to speakers; (c) secondment of research personnel from companies.
(ii) Research contracts and co-operative research are generally not as frequent as the exchange of information based on human contact.

It can be seen from Table 4.1 that the most common form of interchange is advice or consultation about industrial R&D from academics, most importantly from universities (Table 4.1E). On the whole this form on interchange is the most frequent, and represents 17 per cent of all identified associations.

More than half the corporations questioned (51 per cent) had received advice or consultation from universities, and 32 per cent

Table 4.1 Industry–academic interaction rates

	Partner for interaction					
Type of interaction	*Industry*	*University*	*National laboratory*	*Special corporation*	*Foreign countries*	*Total*
A	16	29	13	7	9	11
B	15	3	17	14	2	9
C	23	5	7	4	5	7
D	1	2	11	4	—	3
E	17	51	32	9	10	17
F	33	11	4	4	8	9
G	5	23	6	2	4	6
H	4	4	1	+	4	2
I	9	28	15	4	13	10
J	7	24	13	7	8	9
Average	13	18	13	5	6	8

Source: Science and Technology Agency survey.

Figures are given as percentages.

Notes: A, contract research in which industry asks for research work from partners; B, contract research in which industry is asked for research work by partners; C, contract research in which industry works collaboratively with partners; D, collaboration through government commission research; E, partners providing advice and consultation work to industry; F, exchange of specific knowledge; G, partners providing lectures to industry; H, partners received as research fellow at industry; I, partners accept junior reseach fellows from industry; J, partners accept senior research fellows from industry.

from national laboratories. Twenty-three per cent of the corporations regularly invite speakers from universities (Table 4.1G). Payment for the speaker's services is quite informal and therefore cannot be included in the account of the R&D fund flow from industry to academia. It nevertheless is a great stimulator!

There is substantial human interaction in the form of secondments of research personnel from the company to universities (see I and J in Table 4.1). About 25 per cent of the corporations have a record of posting research personnel to universities, 13 to 15 per cent to national laboratories, and 8 to 13 per cent to foreign countries.

Although it is outside the scope of this analysis, it is noteworthy that there is also a lively research co-operation and contract research between corporations as well (see the 'Industry' heading in Table 4.1).

The second method of assessment is to follow the flow of R&D funds accompanying industrial–academic interaction. Here we utilize the *Report on the Survey of Research and Development* (RSRD) issued by the Prime Minister's Office.

Tables 4.2 and 4.3 indicate the ratios of national and private funds employed by the various research organizations for intramural research expenditure. Table 4.2 contains figures of the R&D funds as a whole (that is, natural science, humanities and social science research funds, taken together), whereas Table 4.3 relates to natural science research funds only.

Table 4.2 Allocation of R&D funding by sector, 1983, Japan (total)

R&D sectors	*No. of research units*	*R&D funding sources* *Government R&D funds*	*(%)*	*Private R&D funds*	*(%)*	*Total*	*(%)*
University[a]							
National	504	697	98	14	2	711	100
Local government	93	78	100	0.3	—	78	100
Private	718	129	15	731	85	860	100
Total	1,315	904	55	745	45	1,649	100
Research laboratory							
National	99	209	100	0.1	0	209	100
Local government	646	191	99	1	1	192	100
Private	301	60	22	219	78	279	100
Total	1,046	460	68	220	32	680	100
Special corporations	9	280	96	11	4	291	100
Industry[b]	17,646	78	2	4,482[b]	98	4,561	100
R&D allocation total	20,016	1,722	24	5,458	76	7,181	100

Source: Prime Minister's Office.

Funding figures are in ¥ billions.

[a] Based on the number of facilities within the universities.

[b] The figure comprises ¥4,475 billion from Japanese domestic industry and ¥6.5 billion from foreign countries.

Of the research funds in national and public universities, 98 to 100 per cent comes from the government, while the corresponding figure for private universities is 14 to 18 per cent, given by the Ministry of Education in the form of subsidies for the costs of setting up research facilities, personnel salaries, and other operating costs.

Of the research funds in national and public laboratories, 99 to 100 comes from the government, whereas the corresponding figure for private laboratories is 22 per cent, derived from the contracts for technology research provided by the government.

Table 4.3 Allocation of R&D funding by sector, 1983, Japan (natural sciences)

R&D sectors	No. of research units	*R&D funding sources* Government R&D funds	(%)	Private R&D funds	(%)	Total	(%)
University[a]							
National	382	547	98	14	2	561	100
Local government	49	49	100	[a]	—	49	100
Private	246	58	14	360	86	418	100
Total	677	654	64	374	36	1,028	100
Research laboratory							
National	96	201	100	[a]	—	201	100
Local government	629	177	99	1	1	178	100
Private	285	53	22	194	78	248	100
Total	1,010	431	69	195	31	627	100
Special corporations	9	277	96	11	4	288	100
Industry	17,648	78	2	4,482[b]	98	4,561	100
Natural sciences R&D fund grand total	19,344	1,440	22	5,062	78	6,504	100

Source: Prime Minister's Office.

Funding figures are in ¥ billions.

[a] Based on the number of facilities within the universities.

[b] The figure comprises ¥4,475 billion from Japanese domestic industry and ¥6.5 billion from foreign countries.

Since the special corporations conduct specific national policy R&D in the areas of nuclear power, space development, and marine development, some 86 per cent of their research funds come from the government.

Up to 98 per cent of the R&D conducted in industry is privately funded, and 2 per cent is funded by the government, which derives from technology research contracts; the amount of government subsidy is negligible. In addition, should patent rights result from such research contracts, these are the property of the government according to law, although the private corporation concerned can obtain a license to use such patents.

Industrial–academic interaction from the perspective of the university receiving R&D funding

The R&D funds received by universities from the private sector derive from designated donations, fees for contract or collaborative research, and donations from private research support corporations. In FY 1983 universities reported to the Prime Minister's Office the receipt of research funds totalling ¥21,888 million. Of this total, the Ministry of Education data indicate the following figures for R&D funds received from industry by national universities under its jurisdiction.

	¥ millions	*number*
Research donation	14,995	23,525
Collaborative research fund	350	56
Commission research fund	302	241
Total	15,647	23,822

This kind of breakdown was unavailable in the case of public and private universities.

It is revealing to compare the ¥21,888 million total of R&D funding received by the universities from industry and the ¥4,561 billion total R&D disbursement by industry overall.

The ratios of the R&D funds received by universities from industry to the total industrial R&D disbursements in the fiscal years 1960, 1966, 1971, 1974, and 1983 were 0.10, 0.17, 0.34, 0.43, 0.30, and 0.42 per cent, respectively. In other words, this ratio, which was on the order of 0.10 per cent in the 1960s, roughly doubled in the 1970s, and peaked at 0.43 per cent in FT 1974. When industry subsequently experienced operational difficulties under the influence of the oil crisis, real R&D disbursement also levelled off, and this ratio was forced down as well. This is evidenced by the 0.30 per cent figure for FY 1977.

Another significant statistic is that the ¥21,888 million total for R&D funding received from industry by universities in FY 1983 was equivalent to 1.3 per cent of the total university intramural research expenditure.

Disposable university R&D funding, defined as the residue of intramural research expenditure when excluding personnel costs and purchase costs for fixed assets such as land and buildings, totalled ¥506.6 billion in FY 1983. The industrial ¥21,888 million contribution is therefore equivalent to 4.3 per cent of this total (Table 4.4).The ratios of industrial R&D fund inflow to universities to this disposable university R&D fund, in the fiscal years 1961,

Table 4.4 University R&D funding received from various sectors

	University		*National*		*Local gov.*		*Private*	
Source	*Funding*	*n*	*Funding*	*n*	*Funding*	*n*	*Funding*	*n*
Government	166,092	1,435	29,826	522	2,917	76	133,349	837
Special corporation	1,422	182	1,025	133	100	9	297	40
Private sector	21,888	671	16,015	396	257	15	5,616	260
Natural science	20,254	573	15,302	303	254	14	4,697	144
Physical	969	79	880	52	4	2	85	6
Engineering	4,880	228	3,672	110	16	3	1,192	64
Agricultural	987	66	868	42	1	2	118	11
Health	13,418	200	8,882	99	233	7	3,302	63
Foreign country	239	20	166	18	—	—	73	13
Total	189,641	2,308	47,032	1,069	3,274	100	139,335	1,150

Source: RSRD, Science and Technology Agency.

Funding figures are in ¥ millions.

Note: n, Number of university faculties receiving R&D funds.

Table 4.5 Trend of private R&D funding inflows into university natural sciences

	Total		*P*		*E*		*A*		*H*	
Year	*Funding*	*R*	*Funding*	*R*	*Funding*	*R*	*Funding*	*R*	*Funding*	*R*
1961	117	0.9	10	0.4	95	2.1	6	0.4	5	0.1
1966	437	0.9	72	0.9	293	1.4	12	0.3	61	0.5
1971	3,032	3.6	217	1.8	1,627	5.0	156	2.0	1,031	3.3
1974	5,559	3.8	504	2.3	1,569	3.3	249	2.1	3,237	4.9
1977	6,182	3.0	419	1.1	1,969	2.9	388	2.4	3,406	4.0
1980	12,640	4.4	718	1.6	2,813	2.8	673	3.0	6,826	6.1
1981	20,254	5.3	969	1.4	4,880	3.7	987	3.6	13,418	8.6

Source: Nomura Research Institute, RSRD.

Funding is expressed in ¥ millions.

Notes: A, agricultural sciences; E, engineering sciences; H, health sciences; P, physical sciences; R, ratio of industrial R&D inflow to university to disposable university R&D fund (unit: %). Disposable university R&D fund = university R&D fund − wages and salaries − expense on lands and buildings.

Figure 4.1 R&D funding flow into and out of universities, Japan (¥ billions)

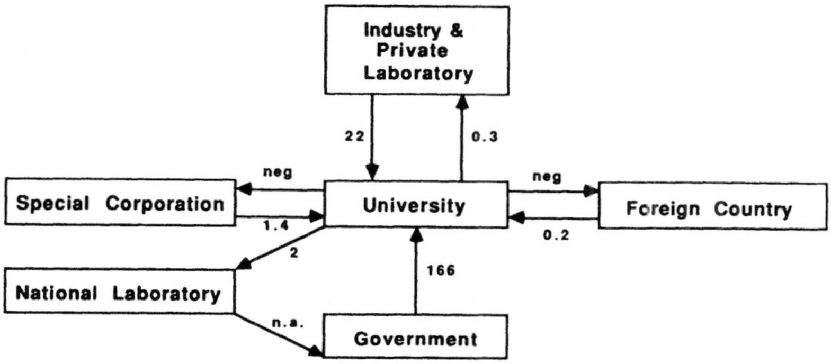

Source: Nomura Research Institute.

1966, 1971, 1974, and 1977, were 0.8, 0.7, 3.3, 3.7, 2.6, and 3.8 per cent, respectively. The trend is thus one of gradual increase, but this is still only a minor contribution. R&D funding flow into and out of universities is summarized graphically in Figure 4.1.

Table 4.5 presents a breakdown of these ratios into four fields of natural science, namely physical science, engineering, agriculture, and the health sciences. One of the longer-term trends is the larger increase in the ratio of private funds in the health sciences. This phenomenon is related to the expansion of pharmaceutical development and new medicines in Japan. In short, the medical departments of universities and the hospitals attached to them are concentrating the development of new pharmaceuticals and the improvement of treatment and diagnostic measurement technology.

A second notable trend is the increase in the above ratios within agricultural science. At the start of the 1980s agriculture drew even with engineering. This is a reflection of the rise in industrial interest in biotechnology.

A third feature is the fact that, while there has been a gradual increase in this ratio in the engineering field, the inflow of private funds into the health sciences has outstripped the corresponding inflow into engineering. For Japan this is a quite unexpected reversal.

Industrial–academic interaction from the perspective of national and public laboratory R&D funding

Table 4.6 indicates the flow of R&D funds received by national laboratories from the private sector and this is summarized in Figure 4.2. These funds consist of fees for research contracts and collaboration, technological guidance, and contract testing, although the quantitative structure is unclear.

Figure 4.2 R&D funding flow into and out of national laboratories, Japan (¥ billions)

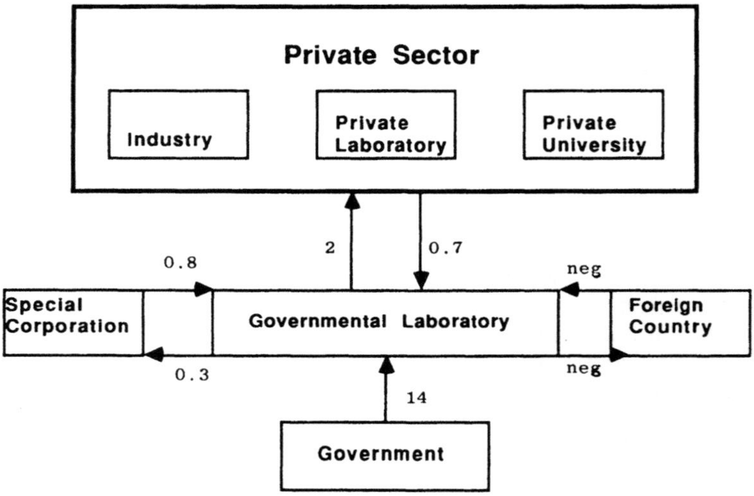

Source: Nomura Research Institute.

When compared with universities, the inflow of private funds is slight. This is especially true in the case of national laboratories, where the total did not even reach ¥100 million. The only substantial inflow was the ¥413 million total received by public laboratories, mainly for the contracts relating to the testing of agricultural chemicals.

In the context of the funding, therefore, the scale of industrial–academic interaction via national and public laboratories is not great. However, this does not necessarily mean that such activity is negligible. This is because the pattern of such interaction surrounding these laboratories revolves around technical consultation (free

Table 4.6 R&D funding flow into and out of national laboratories, Japan

Source	*Inflow to*						*Outflow from*					
	Central government laboratories	*n*	*Local government laboratories*	*n*	*Total*	*n*	*Central government laboratories*	*n*	*Local government laboratories*	*n*	*Total*	*n*
Government	6,588	58	7,402	373	13,990	431	2,537	36	76	48	2,613	84
Special corporation	117	16	717	67	834	83	250	9	6	5	256	14
Private	141	21	598	118	739	139	1,822	26	291	60	2,113	86
Physical	25	2	23	3	48	5	97	1	64	7	161	8
Engineering	64	5	154	17	218	22	1,274	9	104	16	1,378	25
Agricultural	31	10	414	91	445	101	81	10	122	34	203	44
Health	21	3	7	6	28	9	363	5	1	2	364	7
Foreign	3	1	—	—	3	1	*	1	—	—	*	1
Total	6,849	96	8,717	558	15,566	654	4,609	72	373	118	4,982	190

Source: RSRD, Science and Technology Agency.

Funding is expressed in ¥ millions.

Notes: *n*, number of national laboratories concerned; *, positive but negligible.

of charge) and technical guidance (for a fee), the frequency of which is high, but the amount of funds transferred is low. MITI and AIST are making the most vigorous moves in this connection, instituting provisions for contract and collaboration research, and formulating further legislation to promote basic industrial technology. This ministry and its agency are thus attempting to facilitate industrial–academic interaction involving national laboratories, and there is consequently a good possibility that both the frequency and scale of the interaction accompanying the funding flow will increase in the coming years.

Industrial–academic interaction from the perspective of Special Corporation R&D funding

The research activities of the special corporations are funded almost entirely by the government. Still, the Power Reactor and Nuclear Fuel Development Corporation, the Marine Science and Technology Centre, and the Atomic Power Research Institute reported an inflow of funds from the private sector totalling ¥10,648 million.

Table 4.7 R&D funding flows into and out of special corporations, Japan

Source	*Inflow*	*n*	*Outflow*	*n*
Government	343,667	11	423	6
Special corporations	20,686	6	2,571	5
Private	10,832	6	80,377	11
Physical	10,448	3	37,433	5
Engineering	194	1	42,485	2
Agricultural	6	1	0	1
Health	—	—	—	—
Foreign	—	—	829	4
Total	371,185	23	84,200	26

Source: RSRD, Science and Technology Agency.

Funding expressed in ¥ millions.

Note: n, Number of special corporations concerned.

The R&D funding flow associated with special corporations is characterized more by the outflow to industry, universities, and research organs than the inflow of funds they receive. As shown in Table 4.7, and summarized in Figure 4.3, some ¥79,918 million in R&D was disbursed by these corporations to industrial and private research organizations in the form of research contracts and trial

Figure 4.3 R&D funding flow into and out of special corporations, Japan (¥ billions)

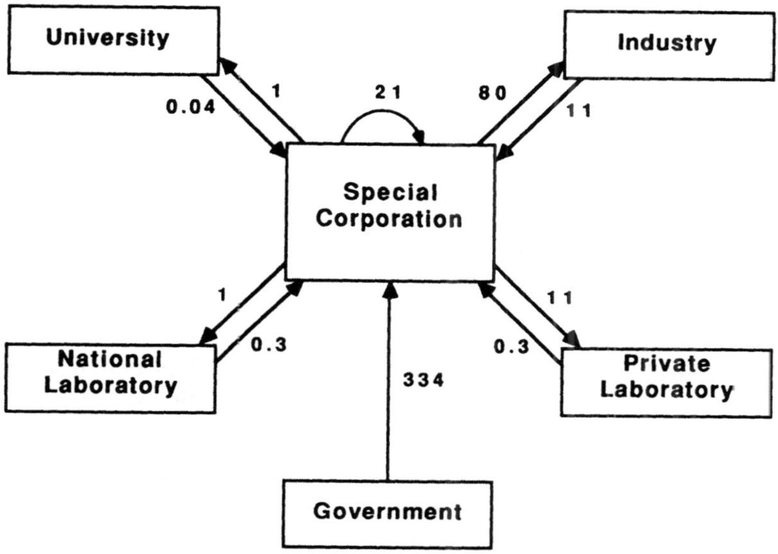

Source: Nomura Research Institute.

manufacture commissions. Research disbursements were also made to universities and national and public laboratories, although these were minor in comparison.

Industrial–academic interaction from the perspective of private laboratory R&D funding

The following account of activities in private laboratories is based on evaluation through interviews, and analysis of ministerial budgetary data (Table 4.8).

First, it is estimated that the bulk of the ¥6.3 billion received by technology research associations from the government is in the form of contracts. The amount may include approximately ¥100 million to ¥1 billion in government subsidies for private laboratories.

Second, there are corporate laboratories operating as independent but affiliated corporations, which receive R&D funds from the parent company as fees for contract research. It is thought that

Table 4.8 R&D funding flows into and out of private laboratories, Japan

Source	*Inflow*	*n*	*Outflow*	*n*
Government	63,053	207	695	66
Special corporations	10,623	111	266	9
Private	197,922	249	15,549	117
Physical	6,202	27	247	13
Engineering	164,214	97	13,797	56
Agricultural	1,609	32	231	14
Health	6,936	34	108	11
Foreign	1,754	13	1,162	13
Total	273,352	580	17,672	205

Source: RSRD, Science and Technology Agency.

Funding expressed in ¥ millions.

Note: n, Number of special corporations concerned.

such disbursement accounts for most of the ¥197.9 billion inflow of R&D funds from the private sector indicated in Table 4.8

Third, there are independent laboratories with the status of public service corporations which receive an inflow of public and private R&D funds. It is difficult to quantify the scale of this inflow. Nevertheless, there are unique examples of industrial–academic interaction within this group which will be outlined later.

Table 4.8 also indicates the outflow of funds from private laboratories. The bulk of the ¥ 695 million outflow to the government is thought to have gone to national universities. A graphical summary of R&D funding flow into and out of private laboratories is given in Figure 4.4.

Industrial–academic interaction from the perspective of industrial R&D funding

Industrial R&D funding accounts for about 76 per cent of total R&D in Japan. It is important, therefore, to consider in what ways funding is channelled from industry to universities and national and public laboratories.

In FY 1983 there was an outflow of ¥198 billion in R&D funds from industry to private laboratories, ¥22 billion to universities and overseas, ¥11 billion to special corporations, and ¥700 million to national and public laboratories, see Figure 4.5.

In the same year it is estimated that industry received ¥69 billion in R&D from special corporations, and ¥56 billion from the

Figure 4.4 R&D funding flow into and out of private laboratories, Japan (¥ billions)

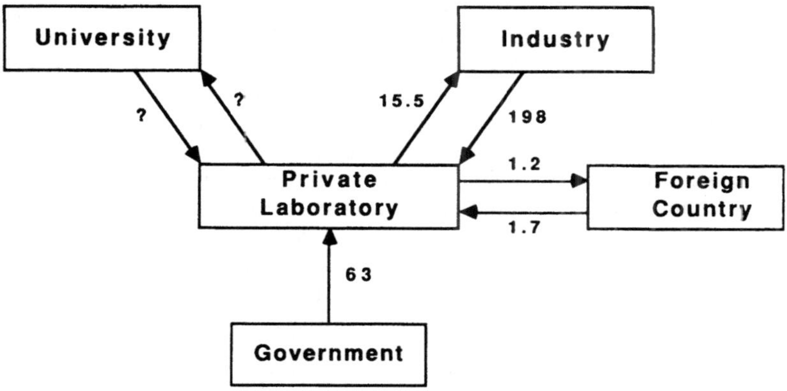

Source: Nomura Research Institute.

Figure 4.5 R&D funding flow into and out of industries, Japan (¥ billions)

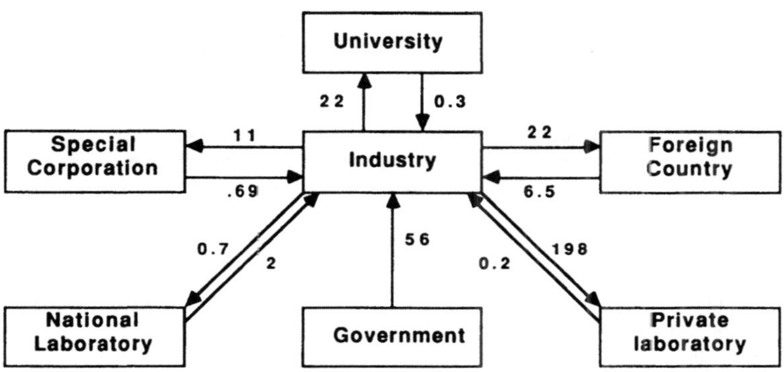

Source: Nomura Research Institute.

government, mainly in the form of fees for contract research. On a more minor scale, there was an R&D inflow totalling ¥6.5 billion from overseas. This was composed mainly of remittances for R&D directed to the Japanese subsidiaries of foreign corporations. In addition, an R&D transfer from national laboratories to industry totalled ¥2 billion.

R&D funding flow from industry to academia

In 1983 Japanese industry disbursed R&D funds totalling ¥16 billion, ¥0.3 billion, and ¥5.6 billion to national, public, and private universities, respectively, a total of ¥21.9 billion. The disbursement of R&D funds from universities to industry, on the other hand, totalled ¥0.3 billion. Details about the flow of R&D funds to universities from industry according to industrial sector are not sufficient to clarify the overall activity. For example, there are no such data about private universities.

Data on the flow of R&D funds to national and public universities from industry by industrial sector is consolidated within the corresponding data for national and public laboratories (Table 4.10). Table 4.9 shows that the inflow of R&D funds from industry to national and local government research organizations (national universities and national/public laboratories) totalled ¥17 billion. Of this, 94 per cent was directed to national universities. Consequently, the data in Table 4.10 can virtually be regarded as depicting the R&D flow to national universities. Although the total ¥15 billion R&D funding from industry indicated in Table 4.10 does not coincide exactly with the ¥17 billion total in Table 4.9, it does give a broad indication of the overall position.

The following points can be noted:

First, the corporations making R&D disbursements to national universities numbered 1,144 for the year. This is equivalent to 6.5 per cent of the 17,646 Japanese corporations implementing R&D,

Table 4.9 R&D funding flows into and out of industries, Japan

Source	*Inflow*	*%*	*Outflow*	*%*
University	302	0.2	21,888	8.6
National	240	0.2	16,015	6.3
Local government	3	0.0	257	0.1
Private	59	0.04	5,616	2.2
Public laboratory	2,113	1.6	739	0.3
National	1,822	1.4	141	0.06
Local government	291	0.2	598	0.2
Special corporation	69,457	51.5	10,832	4.3
Private laboratory	155	0.1	197,922	78.2
Foreign	6,515	4.8	21,819	8.6
Government (central and local)	56,389	41.8	—	—
Total	134,931	100.0	253,200	100.0

Source: RSRD, Science and Technology Agency.

Funding expressed in ¥ millions.

Table 4.10 Outflow of industrial R&D fund by industrial sector

Industrial sector	*(1)*	*(%)*	*(2)*	*(%)*	*(3)*	*(4)*	*(5)*
Agriculture, Forestry and Fishery	+	+	2	+	1	+	+
Mining	17	+	4	+	14	4	0.11
Construction	443	3	94	8	7	5	0.44
Manufacturing	14,332	94	1,027	90	6	14	0.34
Food	1,104	7	180	16	13	6	0.99
Textile	271	2	29	3	3	9	0.53
Pulp and paper	30	+	5	+	2	6	0.12
Printing and publishing	+	+	2	+	4	+	+
Chemical industry	10,380	69	377	33	18	28	1.34
Pharmaceuticals	8,168	54	131	11	30	62	2.82
Others	2,212	15	246	22	15	9	0.46
Petroleum and coal products	310	2	8	1	6	39	0.62
Rubber products	39	+	9	1	5	4	0.07
Glass and ceramics	124	1	41	4	5	3	0.11
Steel	173	1	24	2	16	7	0.09
Non-steel metals	126	1	30	3	13	4	0.17
Fabricated metal products	69	+	33	3	2	2	0.08
Machinery	220	1	82	7	4	3	0.07
Electric machinery	855	6	93	8	4	9	0.06
Electric	344	2	31	3	4	11	0.08
Electronic	511	3	62	5	5	8	0.05
Transportation machinery	71	+	16	1	2	4	0.01
Automobiles	20	+	10	1	5	2	0.00
Others	51	+	6	+	1	8	0.05
Precision machinery	356	2	65	6	9	5	0.22
Others	198	1	33	3	2	6	0.17
Transportation, telecommunication and public utilities industries	405	3	17	1	30	24	0.22
Grand total	15,198	100	1,144	100	6	13	0.33

Source: RSRD, Science and Technology Agency.

Notes:

(1) R&D fund flow to national university (¥ millions).
(2) Number of companies pursuing interactions with national universities.
(3) Industry–university interaction rate, defined as the ratio of number of companies pursuing interactions with national universities to total number of companies executing R&D activity.
(4) Average R&D fund flow (¥ millions) per annum per company pursuing interactions.
(5) The ratio of R&D fund flow to the National Treasury to the total R&D expenditure by industrial sector.
+, Positive but negligible.

and 28 per cent of the 4,062 such corporations capitalized at ¥100 million or over, which group accounts for 95 per cent of Japan's industrial R&D funds. Table 4.11 indicates, for the industry as a whole and the manufacturing industry separately, the ratios of

Table 4.11 Industry–university interaction, according to company

	(1)	(2)	(3)	(4)	(5)	Average
Total of industry	56	21	10	4	4	6
Manufacturing	61	22	10	4	4	6

Notes:

(1) Company capital size more than ¥10 billion.
(2) Company capital size between ¥1 billion and ¥10 billion.
(3) Company capital size between ¥100 million and ¥1 billion.
(4) Company capital size between ¥20 million and ¥100 million.
(5) Company capital size less than ¥10 million.

corporations implementing industrial–academic interaction by capitalization level.

As for the corresponding ratios by industrial sector, the highest are recorded by the pharmaceuticals industry and transportation, telecommunications and public utilities, at 30 per cent (see column (3) in Table 4.10). These are followed by the chemical industry (excluding pharmaceuticals), steelmaking, the non-ferrous metals industry, the mining industry, and the food industry, in which cases some 13 to 16 per cent of the corporations implement industrial–academic interaction. The high ratio for the food industry can be understood as a function of the numerous corporations attempting diversification into the fermentation processes of pharmaceutical industries.

In the machinery industry in the broad sense (general machinery, electrical machinery, transportation machinery, and precision machinery), the ratio is a low 2 to 9 per cent.

Second, it should be noted that the manufacturing industry accounts for 94 per cent of the R&D funds disbursed to universities from industry. Viewed by sector, 54 per cent of this total comes from the pharmaceuticals industry, 15 per cent from the chemical industry (excluding pharmaceuticals), and 7 per cent from the food industry. The chemical industry in the broad sense (including ceramics, rubber products, synthetic textiles, and other such industries in addition to the above accounts for 81 per cent of the total.

In contrast, the flow of R&D funds from the machine industry, in the broad sense, to universities accounts for 10 per cent of the total. This is perhaps due to the following factors: (1) the ratio of developmental research expenditure is high, and that of basic research low; (2) as a consequence, contact with universities and other academic research organizations is not so necessary; (3) R&D is of a more self-sufficient character in the machine industry.

The company average annual R&D disbursements to national universities for corporations capitalized at over ¥10 billion are substantial, with the exception of the automobile industry (Table 4.12). Nevertheless, the contribution still remains less than 9.1 per cent of their total R&D disbursement (Table 4.11 (5)).

Third, Table 4.12 shows the company average disbursements of R&D funds to national universities by capital size, according to sectors of manufacturing industry.

Table 4.12 Company average disbursement of R&D funds to national universities according to company size

	Industrial R&D fund outflow to national university by capital size					
Industry	*(1)*	*(2)*	*(3)*	*(4)*	*(5)*	*Average*
Manufacturing	61	29	7	2	0.5	13
Drug	263	124	25	4	1.3	62
Chemicals	59	13	3	0.8	0.7	9
Petroleum and coal products	75	3	—	—	—	39
Glass and ceramics	9	3	2	1	—	3
Steel	15	5	1	—	—	7
Machinery	16	4	1	1	—	3
Electric	64	8	2	0.7	—	11
Electronics	28	20	1.5	1.2	—	8
Automobiles	3	—	1	—	—	2
Precision machinery	28	8	4	4	—	5

Source: RSRD, Science and Technology Agency.

Units are in ¥ millions per company.

Notes:

(1) Capital size of company more than ¥10 billion.
(2) Capital size of company between ¥1 billion and ¥10 billion.
(3) Capital size of company between ¥100 million and ¥1 billion.
(4) Capital size of company between ¥10 million and ¥100 million.
(5) Capital size of company less than ¥10 million.

Pharmaceuticals capitalized at over ¥10 billion disbursed an average of ¥263 million per company yearly to national universities. The corresponding figure for the 33 pharmaceutical manufacturers capitalized at between ¥1 billion and ¥10 billion was ¥124 million. When private universities are included, the estimated per company averages for pharmaceutical manufacturers capitalized at over ¥10 billion and those in the ¥1 billion to ¥10 billion range are ¥358 million and ¥169 million, respectively.

For the remainder of manufacturing industry there is virtually no disbursement of R&D funds to national universities by corporations capitalized at less than ¥100 million. Even in the case of those capitalized at between ¥1 billion and ¥10 billion, the average annual disbursement per company amounts to a couple of million yen at most.

R&D funding flow from industry to foreign countries

It is estimated that the ¥21,891 million outflow of R&D funds from Japan to foreign countries consists of the following four major categories:

1. R&D fund remittances to overseas laboratories of Japanese corporations
2. Expenditure by Japanese corporations for industrial–academic interaction with US and Western European universities
3. Fees for contract research from Japanese companies to independent contract research institutes overseas
4. R&D funds sent by the Japanese of foreign corporations to the parent company

In the first category, the main Japanese corporations with laboratories overseas are Kirin Brewery (US, California), Otsuka Pharmaceutical (US, Maryland), Mitsubishi Electrical (US, Massachusetts), Honda (US, California, and FRG, Offenbach), Aishin Precision Machinery (France, Cannes), Kobe Steel (US, N. Carolina), and Sumitomo Electric (US, N. Carolina). These were established recently, and the volume of remittances in the outflow of R&D funds overseas in FY 1983 is estimated to be small. Nevertheless, there is a strong possibility that this volume will increase rapidly in the future.

Industrial–academic interaction between Japanese corporations and foreign (US and Western Europe) universities is exemplified mainly by the participation of approximately 50 Japanese corporations in the Industrial Liaison Program (ILP) of the Massachusetts Institute of Technology in the United States. The ILP was set up to provide a venue for industrial–academic interaction in FY 1948. For a $30,000 annual membership fee the corporation is reportedly provided with information on dissertations and lectures by MIT professors, as well as access to seminars and introductions to researchers. In addition, there are about ten Japanese corporations that support MIT professorships. Such professorships require a donation of the order of $100,000 annually.

Another example of such industrial–academic interaction by Japanese corporations is provided by the secondment of research

Table 4.13 Industrial R&D funding outflow to foreign countries from Japan

	(1)		*(2)*		*(3)*			*(4)*		
Industry	*Outflow*	*%*	*No. of companies*	*%*	*A*	*B (%)*	*C (%)*	*A*	*B (%)*	*C (%)*
Agriculture, Forestry, and Fisheries	+	+	1	+	1	+	50	+	—	+
Mining	3,025	14	3	1	10	10	25	1,008	+	+
Construction	82	0.4	65	21	5	+	+	1	—	—
Manufacturing	16,924	79	236	77	1	8	38	73	108	147
Food	55	0.3	7	2	1	7	10	8	+	+
Textile	255	1	15	5	1	—	50	17	+	+
Pulp and paper	+	+	2	0.6	1	—	40	+	+	+
Printing and publishing	+	+	1	+	2	—	50	+	+	+
Chemicals	10,555	40	74	24	4	19	57	143	160	219
Pharmaceuticals	2,931	13	27	9	6	37	70	109	74	262
Others	7,624	35	47	15	3	14	52	+	248	197
Petroleum and coal products	198	1	5	1.6	3	13	22	36	+	+
Rubber products	+	+	2	0.6	1	8	100	+	+	+
Glass and ceramics	241	1	8	2.6	1	6	33	30	21	40
Steel	734	3	10	3	7	4	89	73	+	+
Non-steel metals	593	3	7	2	3	13	14	85	+	+
Fabricated metal products	17	2	7	2	+	+	+	2	—	—
Machinery	138	0.6	10	3	+	5	17	14	+	+
Electric machinery	1,541	7	19	6	1	6	22	81	90	156
Electric	468	2	8	2.6	1	8	13	59	+	+
Electronic	1,074	5	11	3.4	1	4	27	98	110	150
Transportation machinery	2,254	10	15	5	2	7	63	150	49	201
Automobiles	1,909	9	10	3	5	10	60	191	61	278
Others	345	1	5	2	1	+	67	69	+	+
Precision machinery	282	1	10	3	1	4	+	28	+	—
Others	61	0.3	44	14	2	2	+	1	+	—
Transportation, telecommunication, and public utilities	1,416	6	3	1	5	—	13	472	—	475
Grand total	21,447	100	300	100	2	7	32	71	148	155

Source: RSRD, Science and Technology Agency.

Notes:

(1) R&D fund flow to foreign countries (¥ millions).
(2) Number of companies pursuing interactions with foreign countries.
(3) Interaction rate of Japanese industry with foreign countries, defined as the ratio of number of companies pursuing foreign interactions to total number of companies involved in R&D activity.
(4) Average R&D fund flow (¥ millions) per annum per company pursuing foreign interactions.

A, average; B, capital size of company between ¥1 billion and ¥10 billion; C, capital size of company more than ¥10 billion; +, positive but negligible.

personnel to Britain to conduct collaborative research in the development programme at Aberdeen University.

Well-known cases of contract research to independent institutes overseas by Japanese corporations are those with Battelle Research Institute in the United States and West Germany. Japanese corporations also station engineers and scientists in New York, London, and Dusseldorf to monitor research trends in the west and initiate interaction if necessary.

Corporations forwarding R&D funds to foreign countries numbered 308 in FY 1983. This is equivalent to 1.7 per cent of the Japanese corporations engaged in R&D. In the manufacturing sector alone the figure is 1.5 per cent.

The chemical industry (including pharmaceuticals) disbursed the most R&D funds to foreign countries. It is significant also that the automobile industry, which disbursed a negligible amount of such funds to universities in Japan, disbursed a total of ¥1.9 billion for research funds overseas (Table 4.13 (1)).

The ratio of corporations implementing such interaction with foreign countries rises sharply when the capitalization level passes the ¥10 billion mark, averaging 32 per cent of the corporations in the industry as a whole, and 38 per cent in the manufacturing industry (Table 4.13 (3)).

The company average of R&D funds sent overseas annually totalled ¥71 million in the case of the industry as a whole. This figure rises to ¥151 million for corporations capitalized at over ¥1 billion. The corresponding averages for industrial–academic interaction with Japanese national universities in the industry as a whole and for those capitalized at over ¥1 billion (¥13 million and ¥1 billion, respectively) imply a disparity in the manner of conducting the academic liaison.

Britain

There is little detailed information compared with our research in Japan for the support given by British companies to university research or contracts. However, the tables below demonstrate that the sums involved are considerable (Tables 4.14 and 4.15).

Universities and polytechnics receive around £30 million per annum from industry. On average each polytechnic received nearly £80,000 per annum from industry, whereas the comparable figure for universities is over £600,000. These two averages relate to the most recent years for which data are available (1981 and 1982). Not included in these figures are the provision of materials, equipment, and staff time. Also not included are large donations (non-research

Table 4.14 Financing of university links with industry, Britain

Size of direct industry funds annually £	*Universities*
0–100,000	4
100,000–250,000	10
250,000–500,000	15
500,000–750,000	5
750,000–1,000,000	2
1,000,000 and over	9
	45
Total £26.5m (approx.)	
Total sample size = 45 universities	

Source: ACARD study of links between HEIs and industry.

Table 4.15 Financing of polytechnic links with industry, Britain

Size of direct industry funds annually £	*Polytechnics*
0–25,000	6
26,000–50,000	5
51,000–75,000	4
75,000–100,000	2
101,000–150,000	1
151,000–200,000	1
201,000–250,000	3
	22
Total sample size = 22 polytechnics	

Source: ACARD study of links between HEIs and industry.

These data refer to different time periods and are based on possibly inconsistent aggregations.

specific funds) for research laboratories (e.g. the £2.7 million that Sandoz has given to University College London has been excluded).

There are four levels of financial involvement by industry. At the top there are those companies devoting in excess of £1 million to British universities. This includes such firms as ICI, BP, and Shell. Beneath this super league of firms came those devoting in excess of £100,000 per annum (e.g. Pilkingtons, Toche, and Gallahers). The remaining two levels are those companies devoting funds in excess of £25,000 and £6,000 respectively.

The big spenders reported for one specific year are:

	£
BP	3,080,000
ICI	1,800,000
Shell	1,115,000
Pilkingtons	378,000
Pfizer	350,000
Gallaher	265,000
Searle	204,000
Roche	194,000
National Nuclear Corp.	130,000
Lucas	107,000
British Shipbuilders Research	96,000
British Railways	90,000
Dalgety Spiller	85,000
Amersham International	53,000
Blue Circle	40,000
Rank Hovis McDougall	30,000

The following companies also spend over £25,000 p.a.:

Plessey
Micro Consultants
May & Baker
British Gas
ITT
Castrol
Ferranti
Cosworth

These figures make it difficult to infer a great deal about out-of-house R&D expenditure, let alone relations with institutes of higher education. On this subject all that can be stated is that the smaller the firm, the smaller the expenditure and the more likely it is to be in kind rather than cash.

Finally, in assessing the size and the extent of the financial links between universities and industry it is important to remember the following point:

> It is often the case . . . that the industrial research contract which shows up in a university's account is merely 10% of the Company's total investment in a particular project, but that the University-based work is the crucial element which underpins the whole project which would be impossible without it.
>
> (Prof. John Ashworth, University of Salford)

Part III

Case studies of industry — education and co-operation

Chapter five

Case studies: Japan

Industrial initiatives for interaction

Here we examine five case studies in which industry took the initiative in forming the interaction. It does not, therefore, include instances in which industrial corporations used the systems instituted by the MOE or other government bodies.

Sagami Chemical Research Centre

The SCRC is one of the organizations set up in 1963 to commemorate the 60th anniversary of the founding of the Industrial Bank of Japan. It was established through the co-operative efforts of 14 chemical companies and the Industrial Bank of Japan as a venue for basic research linking chemistry and industry. The 14 companies are: Onoda Cement, Kyowa Hakko, Kuraray, Daikyo Oil, Toyo Soda, Nittobo, Mitsubishi Gas and Chemical, Kanegafuchi Chemical, Kyowa Gas and Chemical, Central Glass, Chisso, Nissan Chemical, Japan Mining, Hodogayo Chemical.

The directors managing the SCRC come from industry and academia. The SCRC aims to conduct research oriented toward the discovery and invention of fundamental chemical technology. The technology can be licensed to companies that are not investors in the centre. Since its establishment the centre has accumulated a good research record and has been widely publicized in industry through patent applications and trade papers. It has received many awards from scientific societies. Table 5.1 details the technology that has been licensed to industry.

The SCRC's research record and its contribution to industrial technology has been significant and is mainly due to the flow of talented personnel through the centre. This could be termed the salient point of SCRC's industrial–academic interaction. The following individuals are researchers who compiled excellent

Table 5.1 Sagami Chemical Research Centre — industrial technology development and licensing

Year	Licensed technology
1969	Manufacturing technology for carbonyl sulfide, raw material for herbicide 'Saturn' (Ihara Chemical)
1970 –73	Synthesis technology for DAMN derivatives (Japan Soda, Kyowa Gas and Chemical)
1972	FAMSO manufacturing technology (Japan Soda): formaldehyde dimethyl-mercaptal-*S*-oxide
1973	NO_2 concentration technology (Hitachi, Ltd.)
1974	Novel synthesis of the dihalovinyl analogues of chrysanthemate, a superior insecticide
1976	Manufacturing technology of phenyl acetic acid and thienylacetic acid (Japan Soda)
1978	Novel synthesis of anti-inflammatory analgesics, Diclofenac (Nissan Chemical) and Alclofenac (Japan Soda)
	Measuring method and manufacturing know-how of the measuring instrument for ^3H and ^{14}C
	Dehydration technology for radioactive water (Aroka, Fuji Electrical Machinery)
1979	New method for synthesizing antiphlogistic anodyne 'Tolmetin' (McNeilabs, Dainippon Seiyaku)
1980	Basic technology for synthesis of 'Salmon Calcitonin' (Teikoku Zoki Seiyaku)
	Manufacturing technology for gas analysis equipment (Mitsubishi Yuka)
	Standard solution preparation technology (Wako Junyaku)
1981	New process for industrial production of alpha-arylackanoic acids (Wyntex Inc. — U.S.A.)
	Manufacturing method of Keloprofen (Nissan Chemical)
1982	Manufacturing methods for fluoro-compounds (Onoda Cement, Toyo Soda)
1983	New method of synthesizing aspart-heme and phenylalanine (Toyo Soda)
1984	Manufacturing technology for 1,2,3-thiadiazole derivatives (Japan Redaley)

Source: Sagami Chemical Research Centre.

records in scientific technology at the centre and later became university professors: Prof. Ojima, University of New York (synthesis using organic–metallic compounds); Prof. Tsuchihashi, Keio University (organic synthesis using sulphur compounds); and Prof. Uehara, Josai University (gas analysis equipment for ammonia and other gases). Conversely, the following two individuals are assistant professors who entered the centre from academia and quickly achieved excellent records as researchers: Dr Terajima, from Tokyo University, responsible for development of new manufacuturing methods for anthracyclin; and Dr Shibasaki, from Teikyo University, proposed stereoisomeric synthesis for prostaglandin 12.

There have been virtually no cases of researchers from the centre entering corporations. The centre has instead had a great amount of personnel interaction with academia. There is in addition a system of secondment from private corporations to the SCRC.

As of April 1983 the centre was manned by a total of 155 (93 centre researchers, 19 researchers seconded from corporations, and 43 supporting staff), and had 74 patents pending, and 50 cases of announcements or dissertations at scientific societies.

The FY 1981 budget reportedly totalled ¥1,244 billion. Donations and revenue from patent royalties each accounted for 40 per cent of this total, thus constituting the major sources of revenue. The remainder of the budget is supplied by contract research from the government and the operation of endowments. Even though the centre is an institute created and supported by industry, it cannot expect to rely constantly on donations, given the circumstances of the industrial society in Japan. Therefore an increase in patent revenue is deemed essential for its continued operation. Nevertheless, the individual research results obtained at the centre provide it with a fine record in scientific interest, but the bottom line is that results that do not bring in funding do not contribute to its self-sufficiency. For this reason, the number of researchers peaked in 1978, and the scale of its operations, which had been on the rise until that year, has since declined.

The SCRC system exemplifies personnel and research exchange on the initiative of industry, and an attempt to achieve self-sufficiency through patent revenue.

Toray Science Foundation

The TSF was established through an endowment from Toray Ltd in 1960. At that time Japan lagged considerably behind the United States and Europe in its level of science and technology. Likewise, the national financial capability was weak, and national subsidization of universities and national and public laboratories was extremely limited. In 1960, for example, MOE research subsidies amounted to a mere ¥1,820 million.

Toray consequently set up this foundation to provide financial assistance for basic research in the physical sciences and engineering. Each year the foundation makes disbursements in the form of aid to scientific and technological research (¥100 million; ten at ¥10 million per year) and awards for significant achievements in science and technology (¥10 million; four at ¥2.5 million per year). It also awards commendations to teachers who have rendered outstanding service in scientific education. Its scientific education award is given to middle school and high school teachers (ten at ¥400,000 per year).

TSF was one of the first organizations of this type after the war to begin aid to scientific and technological research. For the times, the scale of its aid was extremely great (¥10 million), and those working in the field of basic research benefited greatly from this help. Including those in existence since the pre-war period, Japan's research assistance organizations currently top 100. The number of foundations formed by corporations or the personal investments of industrialists has recently been climbing. The objectives are diverse, including aid for museums, scholarships, the promotion of science and technology, international exchange, social welfare (for children, women, and the aged), and even sports. The proportion of those carrying out activities to aid research is above 40 per cent.

The scale of these activities is extremely small compared with the Rockefeller or Ford foundations, but even so the yearly disbursements of the 25 major foundations aiding research total an estimated ¥5.4 billion.

The extent and direction of the aid given by industry for research in academia is generally based on selections made by a committee composed of academics. No compensation is expected in return for the aid.

Research aid may be defined as a system of industrial–academic interaction in the form of financial assistance rendered by industry, and TSF is a typical example. Table 5.2 indicates the scale of the funding activities of TSF and other similar foundations.

Hayashibara Forum

The Hayashibara Biochemical Research Institute was detached from its parent company, Hayashibara Ltd, and set up as an affiliated company in 1970. The Hayashibara Group is engaged in

Table 5.2 Selected examples of foundations assisting research

Foundation	*Annual expenditure (¥ millions)*
Toyota	852
Nippon Life Insurance	710
Mitsukoshi Department Store	409
Sanai	322
Sony	300
Kyocera	300
Nissan	294
Honda	260
Takeda	220
Toray	194
Asahi Glass	145

Source: Nomura Research Institute.

a variety of businesses, including real estate leasing, hotels, and sales of food products and pharmaceutical raw materials. It is a group of medium standing, with a yearly turnover of some ¥30 billion. Out of this total, it expends about ¥2 billion on R&D, which is carried out at the research institute.

The operation is distinguished by the forceful leadership of the company president, Mr Hayashibara; the concentration of energies on research fields designated by the president; and the unique bridgeheads it has already established in biochemical research: a method of human cell cultivation in live hamsters, CBG (cancer breaking factor), TNF (tumour necrosis factor), OH-1 (anticancer substance), and malto-pentaose (a diagnostic reagent used for pancreatic diseases).

However, the concern of this book is the system of industrial–academic interaction, and in this chapter, that which arises from the initiative of industry. In the case of the Hayashibara Biochemical Research Institute, this means its sponsorship of the Hayashibara Forum.

In 1985 a forum entitled, 'Life Sciences and Computer Application' was held for four days in July. It was hosted by the Department of Mathematics of Okayama University, which is located in Okayama City, also the site of the institute. Hayashibara sponsored the forum and made available the facilities of the research institute. The forum was attended by both Japanese and foreign scholars, mainly from the United States and Canada.

Hayashibara has also held *Zahlentheorie* symposia and interferon symposia, and is intending to continue doing so in the name of the Hayashibara Forum in the future.

There are other examples of industrial–academic interaction on the initiative of industry in the form of sponsorship of symposia. One such is the Oji Seminar aided by donations made to the Fujiwara Science Foundation, named after its founder Mr Ginjiro Fujiwara, of the Oji Paper Company. The Japan Society for the Promotion of Science (JSPS) is commissioned to select international seminars planned by domestic researchers as subjects for financial support by the company.

Mitsubishi Chemical Life Science Research Laboratory

The MCLSRL was set up to commemorate the 20th anniversary of Mitsubishi Chemical. As a joint-stock company, it naturally has in mind the application of research results over the long term; therefore, the laboratory does not conduct research of a topical (i.e. very short-term) nature. The researchers also determine their own areas of research. These factors characterize the MCLSRL as a basic research laboratory.

The research team of Mitsubishi Chemical Corporation has a 'branch' laboratory in MCLSRL in order to interact with the researchers there. MCLSRL is thus considered a base camp where applied research teams in the parent company look for prospective themes to develop.

There is also a department for overseas interaction, with some foreign researchers in residence for two or three years. Virtually all foreign scholars in the life sciences visiting Japan deliver lectures and pursue discussions at the MCLSRL.

There have been about twenty instances of MCLSRL researchers moving into the academic world as professors or assistant professors. The close interaction with academic circles is further evidenced by the fact that the MCLSRL's vice-director, Prof. Kato, also serves as head of the Japan Embryology Society.

While possession of a basic research department constitutes a financial burden to a company, once it grasps its method of operation the company can unlock the treasures contained therein and make use of them in applied and developmental research. The basic research department is also important for academic interaction. In short, a human bridge is formed through the transfer of researchers to universities and vice versa. In addition active participation in academic activities leads to an exchange of information and scientific thought.

For this reason, large corporations have the margin to fund basic research. Many have recognized the need for such research for the long-term benefit of the company, and have recently begun to establish basic research laboratories. Hitachi Ltd, for example, is setting up a basic research institute in addition to its existing central laboratory.

MCLSRL provides a fine example of a basic research laboratory operated by industry serving as a forum for academic interaction.

Tsukuba Research Consortium

Tsukuba Research Consortium was set up within the Tsukuba Science City as a new type of research organization through the joint efforts of eight corporations (Akashi Manufacturing, Stanley Electric, Teisan, Tokyo Okakogyo, Nippon Heavy Chemical, Japan Vacuum Technology, Hamamatsu Photonics, and Yasukawa Electric). Table 5.3 shows the amount of capital contributed and business nature of the eight companies. TRC has the

Table 5.3 Companies of the Tsukuba Research Consortium

Company	*President*	*Capital (¥ millions)*	*Type of business*
Akashi Manufacturing	K. Akashi	500	Measuring equipment
Hamamatsu Photonics	T. Shoma	450	CdS cell, photoelectronic multiplier, photo tube, TV equipment, etc.
Nippon Heavy Chemical	U. Tumioka	6,800	Ferroalloys, fertilizers, geothermal electric power generation, plant engineering
Nippon Vacuum Technology	C. Hayashi	1,210	Vacuum equipment and engineering
Stanley Electric	T. Kitano	6,200	Automotive electronics components, display devices
Teisan	M. Komatsubara	6,250	Industrial gases and welding equipment
Tokyo Oka Kogyo	T. Ito	470	Photoresistors, electronics, chemicals
Yasukawa Electric	K. Yasukawa	10,580	Electric motors, nuclear equipment, industrial robots, etc.

Source: Nomura Research Institute.

following three aims. The first is collaborative research crossing over the technologies of the eight different corporations.

The second is research interaction between the various national laboratories clustered in Tsukuba Science City. More specifically, this means the provision of a venue for research interaction in the nature of a 'research forum', including the hosting of seminars in advanced technology, science lectures, discussion, and other such events. That is, the consortium does not only furnish the physical facilities such as conference or meeting rooms, but also holds conferences itself (ten to fifteen per year), thus affording a chance for so-called 'soft' interaction. The third aim is research training for young researchers in the eight companies.

The Tsukuba Reseach Consortium was thus founded in the expectation that interaction between corporations in different industries would create a new stimulus. Examples of the themes of collaborative research are conditioning technology to maintain air and water in a super-clean state, three-dimensional display systems for earthquake data, and X-ray microscope lenses. The following are some of the topics of recent lectures held at a forum for research interaction: display technology and its future, Si_3N_4 manufacturing through the CVD method (Tohoku University, Prof. Hirai), vacuum surface technology within nuclear fusion research (Japan Atomic Power Laboratory, Dr Murakami), and ultrasensitive measurement of super-fine particles within the body (Tokyo University, Dr Kawato).

The distinctive feature of industrial–academic interaction in the case of the Tsukuba Research Consortium is the fact that the executives of different corporations decided to set up facilities for collaborative research and make Tsukuba a centre for research interaction. The expectation that interaction between engineers from different industries or researchers in different fields is bound to spawn 'something new' has given rise to numerous 'interaction circles' in other areas.

The forum held once a year by the Science and Technology Promotion Foundation is an example of support by national funds. In these forums researchers from industry and academia in different fields (electronics, biotechnology, and chemistry, for instance) hold discussions on topics in their particular field and in interdisciplinary fields over two days.

There are also many local interaction circles formed by the universities, public laboratories, and industry in almost all prefectures, with the backing of various local commercial and industrial groups and government bodies. These vary in the scale of their activities, but are the same in that they open the door for deeper

Table 5.4 Local industrial–academic interaction circles (selected from more than 50)

City	*Association*
Tohoku	Association for Industrial Technology
Nagano	Interaction Circle
Nagoya	Technology Promotion Forum
Gifu	New Ceramics Circle
Kyoto	Industrial Information Centre
Osaka	Technology Creation Conference
Hiroshima	Industry–Academic Interaction Salon

Source: Nomura Research Institute.

interaction between the individuals involved. Table 5.4 lists some of these circles.

Academic initiatives for interaction

In the second part of this chapter we present four more case studies. This time, however, the initiatives for interaction have come from universities and other academic bodies.

Technology University of Nagaoka

With the notable advances made in technology in recent years, there is a growing need for creative engineers with leadership qualities, who are able to make practical contributions to the common prosperity. With an awareness of such social needs, the National School Establishment Act was amended October, 1976 to permit the founding of this new university. It is characterized by the following four features.

1. The university specializes in engineering but also conducts educational research aimed primarily at practical technological development. It should also be pointed out that some 85 per cent of its graduates go on to postgraduate work.
2. Undergraduates spend about five months in the latter part of the four-year course undergoing practical training at corporations, government agencies, or public corporations.
3. Efforts are made to educate the students in management and social sciences.
4. The university has set up a Technology Development Centre and actively promotes industrial–academic interaction. An

affiliated foundation, the Foundation for Applied Research at the Technological University of Nagaoka (FARTUN), was set up to promote this centre.

FARTUN was founded in 1982, and promotes industrial–academic interaction in the following ways: sponsorship of lectures and research meetings, exchange of information through research reports and trade papers, factory field trips, consultancy services for industry, aid for the acquisition of industrial patent rights, and industrial–academic collaborative research assistance. To aid it in these operations, FARTUN has support from 70 local companies. In addition, the list of those contributing to its endowment exceeds 80 names, and includes local government bodies, representatives of industry, and university teachers.

FARTUN publishes a guidebook titled *Research Conspectus*, which contains the research records of professors and other personnel in the Technological University of Nagaoka. This publication acts to spur industrial–academic interaction.

Since about one-third of the university professors and assistant professors were originally researchers for private corporations, the faculty is well versed in the corporate situation, and this makes for smoother interaction. This active approach has spawned commensurate results.

Table 5.5 indicates that the inflow of funds from industry has risen dramatically in the last few years. Especially notable is the increase in cases of collaborative research and scholarship donations.

The Technological University of Nagaoka is staffed by 62 professors, 65 assistant professors, and 50 supporting personnel, a total of 177. The FY 1983 inflow of funds from the private sector totalled ¥139 million, this exceeding the ¥103 million received as research grants from the MOE. Government funding decreased in FY 1984, while that received from the private sector increased, and is nearing a figure double that from national funding.

The trend exhibited by the Technological University of Nagaoka demonstrates the results of promoting active industrial– academic interaction within a newly established engineering school.

Tohoku University

Tohoku University was established in 1907. It is the third oldest of Japan's universities and is composed of four humanities and social science departments, six natural science departments, and eight research institutes. The following will serve to introduce three

Table 5.5 Inflow of private R&D funds into the Technical University of Nagaoka (¥ millions)

	R&D fund inflow from industry									
Year	*Contract research*	*Cases*	*Collaborative Research*	*Cases*	*Research donation*	*Cases*	*Total*	*Cases*	*MOE research grant*	*Cases*
1978	3	6	—	—	1	2	4	8	2	3
1979	3	6	—	—	20	19	22	25	18	14
1980	11	7	—	—	42	45	53	52	77	21
1981	6	3	—	—	58	77	63	80	69	37
1982	10	8	—	—	89	109	100	117	88	45
1983	16	6	17	4	106	118	139	128	103	59
1984	16	6	30	7	122	158	168	171	87	49

Source: Technical University of Nagaoka.

examples of the unique role this university has played in industrial–academic interaction.

Prof. Honda and Prof. Masumoto — The Research Institute of Electromagnetic Alloys

In 1914 Prof. Kotaro Honda developed the magnetic metal material KS steel. This material had a magnetic resistance of 250 oersteds, over 3.5 times that of the 70-oersted resistance of the tungsten steel common in the west at the time. This important invention, the fruit of basic research on steel and alloys, convinced Prof. Honda of the indispensibility of rigorous basic research for the establishment of Japan's industrial foundation. He consequently proceeded to set up a research institute with the president of Tohoku University, Prof. Tokiatsu Hojo, with funds received from the Sumitomo family. The research institute was renamed the Steel Research Institute, and later the Metallic Materials Research Institute (MMRI).

In 1944 the Research Institute of Electromagnetic Alloys was set up as an affiliated foundation of the MMRI. The research at this laboratory concerns mainly magnetic alloys, and the laboratory receives donations and research consignments from industry. Its current head, Prof. Masumoto, was formerly the head of the Metallic Materials Research Institute, and is a professor emeritus of Tohoku University. Prof. Masumoto has compiled an impressive record, including the development of the magnetic alloys Elinvar and Elcolloy (which had accounted for 90 per cent of Japanese-made watch hair-springs before the digital electronic watch emerged) and Stainless Invar. He is also the discoverer of Matsumoto's Law of magnetic alloy design.

The laboratory is also equipped with a research student system, whereby research students are accepted from the private sector, with many of the results obtained in this connection transferred to industry.

Lured by this record, virtually all of Japan's manufacturers of magnetic materials, magnetic discs and other equipment and components containing magnetic materials make use of this laboratory. This constitutes a special case of interaction on the initiative of academia, triggered by an outstanding record in science and technology.

The revenue of the research institute came to ¥172 million in FY 1982, most of it as fees for trial manufacture and research contracts received from 50 corporations, including watch manufacturers, electronic component manufacturers, wireless communication equipment manufacturers, and general electrical machinery manufacturers.

The institute also receives research students from seven private companies: Sony, Motoyama, Tohoku Special Steel, Matsushita Electric, Yokohama Magnetics, and Alps Electric, as well as professors and senior researchers from Akita University, Tohoku Gakuin University, and Yamagata Prefecture public laboratory. Although the laboratory is staffed by only twenty researchers and sixteen technicians, its record has exerted a powerful attraction for industry.

Prof. Masumoto of the Metallic Materials Research Institute is a son of Prof. Masumoto who heads the Research Institute of Electromagetic Alloys. The former also has an impressive research record, and the following four themes have been adopted from this institute by the RDCJ with a view to industrial application. The names in parentheses indicate the corporations receiving the contracts: manufacturing technology for amorphous metals used as electromagnetic materials (Hitachi Ltd, Hitachi Metals); application technology for high magnetic penetration ratio materials and amorphous metals (Matsushita Electric); application technology for high magnetic deflection materials and amorphous materials (Sony); manufacturing methods for amorphous metals for use in power transformers and iron cores (Nippon Steel).

Furthermore, the Creative Science and Technology Promotion System has formed the Masumoto Structure Material Project as one of its seven themes. It has already been mentioned in the account of the RDCJ activities that it is providing ¥2 billion in backing for this research over a five-year period (Chapter 3, p. 70).

It can be seen from the above that the academic initiatives arising from the combined efforts of the Metallic Materials Research Institute and the Research Institute of Electromagnetic Alloys enhance the role played by Tohoku University in industrial–academic interaction.

Prof. Yagi — The Telecommunication Laboratory — and Prof. Nishizawa — The Semiconductor Research Institute

One of the significant achievements of the Department of Science and Engineering at Tohoku University was the invention of the Yagi antenna by Prof. Yagi. The research on short-wave, long-range radio waves begun by Prof. Yagi in the 1920s was transformed by his successor, Prof. Uta, into ultra-short-wave research. This research, in turn, culminated in the practical utilization of super-short-wave circuits for public use.

Another of Prof. Yagi's students, Prof. Okabe, performed important research on ultra-short-waves generated through split-anode magnetrons. This chain of excellent research concerning short-wave, long-range circuits constitutes one of the major reasons behind the addition of the Telecommunications Laboratory to Tohoku University.

It can be mentioned in passing that Prof. Hideji Yagi subsequently became president of Osaka University, president of the National Technology Agency, a member of the Upper House, and president of the Yagi Antenna Company.

The current head of the Telecommunications Laboratory is Prof. Junichi Mishizawa. In 1961, Prof. Nishizawa donated the industrial rights concerning his invention made in collaboration with Prof. Nei Watanabe, and set up, with the co-operation of industry, the Semiconductor Research Institute.

Prof. Nishizawa built up an outstanding record in the areas of optical electronics and semiconductor engineering. The foundation he set up also gained worldwide attention for the following research efforts: the proposal and development of a static electricity induction transistor (SIT), and three major innovations in the optical communications field: the semiconductor implantation laser, convergent fibres, and APD and p-i-n photo diodes. The work of this institute has been much acclaimed, not only in Japan but also abroad.

Given the pioneering nature of its work, the Semiconductor Research Institute laboured under great financial difficulties early in its history. Thanks to the reputation of Prof. Nishizawa for innovative research, however, many corporations have come into contact with it. Supporting membership presently numbers 31 electrical machinery and electronic materials and components manufacturers. The institute reported a revenue of ¥221 million in 1980. Most of this derived from supporting membership fees, research contracts, funds for research students, and new member fees. Despite the ownership of excellent patents, revenue was slight, reportedly due to mishandling of contracting. This will no doubt be a major topic for future revision. The following examples are patents with practical applications: the photo coupler, the photogenerating diode (super bright red photogenerating diode technology (GaAlAs) based on Prof. Nishizawa's method of vapour pressure control thermal differential liquid-phase growth (to Stanley Electric KK through RDCJ), acoustic SIT (to Nippon Gakki through RDCJ), the ion implantation p-i-n diode (directly to SRI members).

The Institute is hoping to make patent revenue the heart of its operation in the coming years. This aim has surfaced as a desire to conduct development contracts from manufacturers through RDCJ. Based on this policy, the institute has the following development contracts (three cases concern SIT, photogenerating diodes, and one pattern generators, for a total of six). The names in parentheses indicate the contracting corporation: successive manufacturing method for the photogenerating diode (GaAlAs) (Stanley Electric), Audio transistor manufacturing technology (Nippon Gakki), super precision pattern generator manufacturing technology (Kokusai Electric), monolithic display element manufacturing technology (Oki Electric), high-output ultrasonic wave generation equipment manufacturing technology (Tohoku Metal Industries), microwave high-power transistor manufacturing technology (Mitsubishi Electric).

The RDCJ has inaugurated the Nishizawa Perfect Crystal Project as one of the seven themes of the Creative Science and Technology Promotion System, and made Prof. Nishizawa the project team leader. The project was launched in 1981, with a total research budget of ¥2 billion over its five-year term.

In the view of R. W. Noyes, vice-chairman of Intel, SIT was one of the two major inventions of the 1970s. It is Dr Noyes' opinion that it constitutes a development of engineering principles that will have important ramifications for electric power technology.

Prof. Nishizawa's whole-hearted acceptance of research contracted from industry is grounded in his contention that 'basic research can be well activated through application practice'. It is this thinking which lies behind SRI's role of building bridges between academia and industry, and between basic and applied research.

The two institutes attached to Tohoku University and the affiliated research laboratories introduced above all represent cases in which outstanding efforts of university professors in basic or applied research powered the initiative for industrial–academic interaction.

Such interaction at Tohoku University has had the following three effects. The first is the birth of several corporations based on inventions. For example Tohoku Metal Industries (est. 1983, metallic, magnetic materials), Sendai Precision Materials Laboratory (est. 1958, hair spring Elcolloy manufacturing), and Yagi Antenna (est. 1952, antennas).

The second is the siting of factories and laboratories in the vicinity of Tohoku University. Many companies, drawn by the fine achievements of the school, have located factories and laboratories in the area. These include Tohoku Special Steel (magnetic steel), TDK (ferrite magnetic tape), Tohoku Metal Industries (magnetic materials), Sendai Precision Materials Laboratory (Elcolloy), Oki Electric (VLSI), and Kokusai Denki (telecom equipment).

The third is a further intensification of industrial–academic interaction to better effect in Tohoku University. This can also be gauged by the extremely large number of donations made to the university. The school accounts for roughly six per cent of all the donations made to national universities in Japan (6.6 per cent on a monetary basis). Likewise, it is also implied by the large number of themes selected from Tohoku University (mainly the two research organs) by the RDCJ.

Aoba Engineering Research Promotion Centre

This centre has been set up in Tohoku University to aid industrial–academic interaction. The president of the centre is Prof. Ishida, head of the university, while Prof. Musada, the head of the engineering department, is the managing director. Miyagi Prefecture officials also participate in this centre, and the vice-governor is on the board of directors.

The aims of the centre are as follows: (1) to provide financial assistance for the R&D of Tohoku University's engineering department; (2) to arrange sponsorship of lectures and study groups concerning industrial technology; (3) to encourage industrial–academic interaction; and (4) to upgrade the technological level of local industry.

There are at least two other instances of industrial–academic co-operation in the vicinity of Tohoku University: Tohoku Industrial Technology Development Association and Japan Fine Ceramics.

This trend by universities toward furthering industrial–academic interaction and contributing to the technological level of the local industry has begun to surface in national, public, and private universities across the country.

Kitasato Institute

Kitasato is a private institute set up in 1914 by Prof. Shibasaburo Kitasato, who discovered tetanus antibodies and developed a blood serum method of treatment.

Generally speaking, operation of a private institute is an extremely difficult proposition in Japan. However, the operation of the Kitasato Institute has been stable compared with other private institutes since it has been able to gain sufficient revenue through its attached hospital and blood serum business at its manufacturing division. It also has affiliated organizations, such as Kitasato University, Kitasato Junior College of Hygienics, and the Oriental Medical Institute. Each of these is financially self-supporting.

The institute has continued to research blood serum treatment since the days of Prof. Kitasato, but it has also begun research on biochemistry and natural physiologically active substances. At the present time the institute conducts research in virology, antibiotics, immunology, cancer chemotherapy, and the physiological properties of herbs.

The institute has compiled a fine research record in the post-war period and has made a contribution to actual medical treatment. The two foremost examples are the antibiotic leucomycin (kitasamycin) and the cancer-fighting agent mytomycin. Both of these, while rudimentary, are used the world over. They are licensed to Toyo Jozo and Kyowa Hakko Kogyo, respectively. The institute also manufactures vaccines for humans and animals, and drugs used in diagnosis, which are sold through its subsidiary Kitasato Pharmaceutical Industries.

The Kitasato Institute represents a rare case of a private laboratory established on the initiative of academia, which nevertheless accepts research students from private corporations for industrial–academic interaction. The institute licences the results to private corporations, and at the same time engages in the manufacture of medical products to be sold through a subsidiary, all while continuing with its research work.

The scale of its 1981 budget was ¥8,540 million. Of its 801 employees, roughly 60 were classified as researchers.

There are at least ten other similar research institutes, founded by university professors who achieved outstanding results both from a scientific perspective and from practical viewpoints. Three are listed below.

The Institute of Microbial Chemistry was founded by Prof. Hamao Umezawa, formerly of the National Institute of Health. Prof. Umezawa developed a succession of new antibiotics, such as kanamycin (1957), pentamycin (1958), kasugamycin (1959), and bleomycin (1962). All of these were licensed to industry and utilized worldwide. The IMC operates on a yearly budget of approximately ¥800 million, most of it derived from patent revenue and fees for research consignments.

Like the Kitazato Institute, the Chemo-sero Therapeutic Research Institute provides for operational stability through the possession of its manufacturing department and diagnosis department in addition to its research department.

The managing director of the Research Institute of Production and Development (RIPD) is Prof. Seizo Okamura, professor emeritus of Kyoto University. It is a medium sized laboratory with 177 employees, 162 of whom are researchers, serving industry on a contract research and testing basis. It has a hard and fast rule not to accept contracts on the same theme from more than one company. Its FY 1982 budget came to ¥715 million. RIPD covers ten research fields, including food products, microbiotics, medicine, petrochemicals, pharmacology, and strong magnetic materials. The establishment and continued subsistence of RIPD is said to have been greatly aided by the success in developing a manufacturing method of arinamin (a vitamin B_1 derivative) by Prof. Fujiwara of Kyoto University. Arinamin has been manufactured and marketed by Takeda Pharmaceuticals with the patent royalty paid to RIPD.

In each of the above cases, the institute was started on the basis of industrial application of research results obtained by noted academics. The patent revenues from industry and other funds are used as a basis for continuing the research.

A somewhat different case, although no less interesting, is that of the Research Laboratory of Resources Utilization (LORU) attached to Tokyo Institute of Technology. LORU was founded by Prof. Kato, famous not only for his invention of ferrite, but also for his significant achievements in a variety of other chemical fields. He endowed the whole royalty revenue from Nitto Chemical Industries to set up LORU as an attached Laboratory to Tokyo Institute of Technology. He wished to generate industrial–academic interaction through this establishment. Prof. Kato's legacy in this context continues to operate in this Institute today.

Society of Polymer Science

This is a large society, with a formal membership of 8,106, a student membership of 993, and a supporting membership of 346, as well as 88 corporate backers. It is not only a venue for the announcement of academic research results, but also constitutes a site for the study of the interaction of its members, and for proposals concerning national polymer science and technology policy to the government.

In the selection of polymer materials–related themes for the next-generation industrial technology system treated in Chapter 3 (p. 75), a major role was played by the Polymer Club set up within the society. Moreover, the very establishment of this system was assisted by the Polymer Club.

The Society of Polymer Science hosts the Annual Polymer

Science Meeting, the Polymer Discussion Meeting (each once a year), lectures (four times a year), and 'microsymposia' (11 times a year). It has also set up the following conferences to promote industrial–academic interaction with regard to polymer science and technology.

Polymer Summer University

This consists of lectures and discussion meetings held during the summer for four days and three nights concerning the polymer industry and polymer science and technology. There are lectures by university professors, and by engineers and researchers from the user industry, followed by discussions. The substance of the meeting varies, and includes such things as discussions on the future course of polymer science and technology in specific fields, and the exposition of successful R&D cases.

Lectures for pioneering application of polymers

These are held six times a year. Lectures are delivered in subdivisions of specific themes or fields by representatives of both industry and academia concerning the progress of polymer science and technology and the prospects for a wider range of polymer application. These are followed by discussions.

Research groups

Research groups have been formed around the following themes: medical polymers, printing materials, NMR, ultimate materials, polymers and chelate, polymers and water, polymer collapse and stabilization, polymer surface, polymer membrane for separating operations, biopolymers, adhesion and coating, fibres, effective use of carbon resources, electrical properties and static electricity, civil engineering and construction materials, plastic film, molecular weight and its distribution, inorganic polymers, rheology, and plastic materials engineering.

Polymer Club

The Polymer Club was set up mainly by the corporate members of the society. The club debates a variety of topics related to the polymer industry and makes policy proposals to the government. The club holds monthly meetings, and conducts study groups, R&D management research groups, technology management policy study groups, and others. Executives from the 56 member corporations participate in the club's activities, and guest speakers are brought in from universities, national laboratories, industry, and government, making for a lively interaction.

The Society of Polymer Science is not the only example of the formation of a venue for interaction with industry by an academic society.

For the development of biology and biotechnology, for instance, the Fermentation Industrial Society set up the Bio-industry Development Consortium. This group holds courses for the training of researchers, publishes investigative reports (*BIDEC News*), conducts technological and economic studies on the bio-industry, and carries out collaborative and contract research.

The Institute of Electronics and Communication Engineers of Japan is another large society, with a membership of some 32,000. Groups on a similar scale include the Japan Electrical Society (23,000 members) and the Applied Physics Society (21,000 members). They resemble the Society of Polymer Science in that they encourage the participation of representatives of both academia and industry to hold research meetings, courses, and summer universities.

It can also be mentioned here that the total number of such societies registered with the Japan Science Council is 1,178. The following is a list of only the largest, with membership shown in parentheses:

Japan Medical Society (273,000)
Japan Machinery Society (32,000)
Japan Chemistry Society (31,000)
Japan Construction Society (29,000)
Japan Civil Engineering Society (23,000)
Information Processing Society (14,000)
Japan Pharmacology Society (12,000)
Japan Physics Society (11,000)
Japan Steel Association (9,000)
Japan Agricultural Chemistry Society (8,000)

Besides researchers from universities and national, public, and private laboratories, these societies include many corporate members, and their activities are conducted by industry and academia jointly. It can be seen from the above examples that the society activities do not stop at the publication of magazines announcing research results, but also extend to individual research meetings for separate areas of research, and courses for junior researchers. As a result it can be said that these function effectively as vehicles for co-operation between industry and academia.

Chapter six

Case studies: Britain

The following five case studies illustrate various aspects of industrial–academic interaction: Cambridge Science Park, Heriot-Watt Research Park, the Alvey Programme, Salford University Industrial Company, and NEWTECH (Clwyd) Ltd. The Chapter is concluded by a profile of science parks, research parks, and innovation centres.

The Cambridge Science Park

CSP was opened in 1973 and was one of the first science parks to be developed in Britain, and therefore because of its maturity it provides a valuable case study for assessing the role of science parks more generally in fostering local and national economic development objectives.

General background

The idea of a science park in Cambridge evolved over several years and resulted because of the influence of several organizations. Initially in the late 1960s three factors were important:

1. Central government was becoming more concerned as industrial competitiveness in Britain continued to lag behind that in other countries. It was keen therefore to ensure greater contact between institutions of higher education and industry in the hope that such contact would improve product and process development.
2. Cambridge University was becoming increasingly aware of the benefits of interaction with both local and national industry. For some years the university had fostered links with local firms but was keen to retain and strengthen opportunities by advocating that appropriate science-based industry should be permitted to locate in the Cambridge area.

3. A number of groups and organizations in and around Cambridge, including the district councils, were concerned that the area should attract more industry. From the late 1940s onwards industrial development in the Cambridge area was restricted in order that the city should retain its character as a university town.

Against this background the Cambridgeshire County Council, which dealt with local employment policies, accepted the case for relaxing the policies on industry, and amended them in 1971 to permit development of science-based industry. To facilitate this, following negotiations with Trinity College, a science park was proposed and approved on the northeast side of Cambridge.

A legal agreement between Cambridgeshire County Council and Trinity College was signed in November 1971 under section 37 of the 1962 Town and Country Planning Act which specifically reserved the area 'for exclusive use as a science park', that is for

(i) industries closely related to the work of the University which have a high research and design content and which are likely to use the existing or proposed facilities of the University;
(ii) industries where the initial components are so closely integrated that they require control by the same executive personnel throughout the entire process;
(iii) industries which include those processes which precede the manufacturing stages and are activities which are relevant to a continuum of research, design and development leading to manufacturing and marketing, but do not involve either high tonnage productions or mass production, but which employ labour consisting of a high proportion of scientists, technologists, technicians and craftsmen.

The agreement restricted uses appropriate to a science park as:

(i) scientific research associated with industrial production;
(ii) light industrial production of a kind which is dependent on regular consultation with either or both of the following:
  (a) their tenants' own research development and design staff established in the Cambridge Study Area
  (b) the scientific staff or facilities of the University or of local scientific institutions
(iii) ancillary buildings and works appropriate to the use of the said land as a science park.

The nature of industrial development on the science park

Two years after this agreement was signed the first firm moved to the CSP. Table 6.1 shows the subsequent experience. Perhaps what stands out is that despite the undoubted enthusiasm and commitment of the various parties involved the rate of growth of the CSP was quite modest. Seven years after the agreement, only seven firms had opened in the science park. Thereafter, the expansion was more rapid, although by the beginning of the 1980s the number of new openings had fallen back to two per annum. It was not until September 1983 that the science park experienced its first casualty.

Table 6.1 Growth of the Cambridge Science Park, 1973–83

	Number of openings[a]		*Floorspace (000 sq.ft.)*[b]		*Area of development (acres)*	
Year	*New*	*Cumulative*	*New*	*Cumulative*	*New*	*Cumulative*
1973	1	1	15	15	14	14
1974	1	2	25	40	—	14
1975	2	4	11	51	14	28
1976	—	4	—	51	—	28
1977	1	5	16	67	1	28
1978	2	7	4	71	—	28
1979	6	13	53	124	28	56
1980	3	16	14	138	—	56
1981	5	21	17	155	1	56
1982	2	23	15	170	26	82
1983	2	25	124	294	—	82

Source: Moore and Spires (1983).

[a] These are the main companies established on the park. There have been some very small subsidiaries or associate companies opening, but these are excluded.

[b] Approximate figures for each year.

As Table 6.2 shows the firms opening on the science park were concentrated in three main industries — chemicals, instrument engineering, and electrical engineering. This is, perhaps, not surprising given the terms of the agreement when the CSP was established.

Chemicals is the dominant activity but this is partly the result of one very large company (now employing over 200 people) opening at the science park. This industry, although most heavily represented, has the lowest number of employees per 1,000 square feet (2.5), which is less than half that of electrical engineering (5.5).

In general the companies opening at the science park are small. Only two companies employ over a hundred people. Table 6.3

Table 6.2 Type of industry established at the Cambridge Science Park

Industry	*Number of companies*	*%*	*Number of employees*	*%*	*Floorspace area (000s sq. ft.)*	*%*
Chemical	8	32	430	43	171.5	58
Instrument engineering	5	20	190	19	46.5	16
Electrical engineering	5	20	110	11	20.5	7
Miscellaneous manufacturing	1	4	30	3	7.0	2
Business services	3	12	30	3	7.0	2
Professional services	3	12	200	20	41.5	14
Total	25	100	990	100	294	100

*Source:*Moore and Spires (1983).

Table 6.3 Size and structure of the companies at the Cambridge Science Park

	Employees		*Floorspace*		*Age (years)*	
Industry	*Mean*	*Range*	*Mean*	*Range*	*Mean*	*Range*
Chemicals	50	10–230	21.5	3.0–123.0	14	2–44
Instrument engineering	40	10–80	9.0	2.0–25.0	9	3–14
Electrical engineering	20	10–40	4.0	1.5–12.0	7	1–14
Miscellaneous manufacturing	30	30	7.0	7.0	37	37
Business services	10	10–20	2.0	1.5–4.5	7	3–19
Professional services	70	10–180	14.0	1.5–32.5	10	4–23
Total	40	10–230	12.0	1.5–123.0	12	1–44

Source: Moore and Spires (1983).

shows that the size range is indeed quite wide, which is not surprising given the significant differences in both the age of the companies and their length of time operating at the science park.

Of the 1,000 employees at the science park 60 per cent are managerial or professional and technical and only 22 per cent are in the skilled or unskilled manual occupations (Table 6.4).

The majority of the more skilled workers (about three-quarters) were recruited from outside the local area whereas the less skilled

Table 6.4 Occupational structure of employees at the Cambridge Science Park and the percentage of each group recruited locally

Occupational group	*%*	*% recruited locally from the sub-region*
Managerial/professional	23	24
Scientific/technical	37	28
Administration	19	75
Skilled manual	6	59
Semi-skilled manual	9	64
Unskilled manual	6	85
Total	100	46

Source: Moore and Spires (1983).

manual workers tended to be recruited locally. This pattern conforms fairly closely to what is generally found in local labour market studies, namely that the lower the occupational group in the skill hierarchy, the more narrow the geographical delineation of the relevant local labour market.

Finally, it is clear that many of the companies operating at the science park are engaged partly or wholly in R&D and/or other activities embracing high technology. Table 6.5 shows that only three of the 25 companies were not involved in some high-technology activity and only nine were not involved in R&D.

Table 6.5 High technology and R&D activity at the Cambridge Science Park, 1983

Activity	*Number of companies*
High-technology	22
Non-high-technology	3
R&D	16
Non-R&D	9
Total	25

Source: Moore and Spires (1983).

What clearly emerges from this brief description of the type of companies and their activities at the CSP is that they display many of the characteristics identified in the early conception of the science park. Thus industry is heavily skewed towards the science-based sectors. There is significant activity in R&D and high-technology developments, and the occupation structure is biased towards the professional, scientific, and technical groups. However, an important question that remains unanswered is whether this type of development would otherwise have occurred in Cambridge had the science park not been designated and established. This is clearly important from the perspective of those concerned with local economic development and particularly in the case of Cambridge, where development appropriate to an historic university town was considered desirable.

While these early objectives for science parks remain, local authorities have placed increasing importance on the need to improve local employment opportunities. First by providing premises for local firms to start up or expand and second by encouraging the inward movement of firms.

Has the CSP been successful?

From what has been said above, the objectives of the CSP are apparent. First it was established as a vehicle for cultivating closer interaction between research activities in the university and at other research establishments in the area, including R&D going on in industry. This objective certainly figured most strongly in the minds of those who initiated the project. Second, however, emerging as a key objective somewhat later is the notion of the CSP as a means of bringing more jobs into the Cambridge area. In this respect the CSP provides opportunities for local companies to start up and for existing companies to expand, but also acts as a catalyst, encouraging a particular type of industrial development appropriate to an area like Cambridge.

In 1983 Moore and Spires undertook a survey of companies in the science park with a view to analysing the experience of the CSP and assessing whether the above objectives had been achieved. However, in order to disentangle the characteristics and achievements of companies in the science park from companies more generally, Moore and Spires extended the survey to cover all manufacturing companies within the Cambridge sub-region. Hitherto most analyses have focused solely on the experience of science park companies without a 'control group' of companies against which to compare and contrast. The results and conclusions of Moore and Spires survey are detailed below.

Research links

There is little doubt that some companies in the CSP enjoy research links with Cambridge University. However, the survey of CSP companies reveals that only a minority (albeit a significant one) of companies have research links. Table 6.6. shows that about a quarter of all the companies on the CSP have research links with the university. This pattern was similar for both high-tech and R&D firms, although there is, as might be expected, a tendency for more links in the case of R&D activities.

Table 6.6 Incidence and types of research links with the university of companies at the Cambridge Science Park

	CSP			*Cambridge sub-region*		
Links	*All*[a]	*High-tech*[b]	*R&D*[c]	*All*	*High-tech*	*R&D*
Links with the university	24	23	37	19	32	33
Types of links (all companies)						
Exchanged ideas	20	18	31	15	28	25
Exchanged people	4	5	6	—	—	—
Shared equipment	8	5	13	9	11	15
Shared information	4	5	6	11	19	19

Source: Moore and Spires (1983).
Figures are percentages.
[a] 25 companies.
[b] 22 companies.
[c] 16 companies.

What is perhaps surprising is that of the companies operating in the Cambridge sub-region a very similar proportion of those in the CSP claim research links with the university. In other words location at the science park with close proximity to university departments has not encouraged a significantly greater proportion of firms to develop research links than is the case for firms outside the science park. Indeed, firms not in the CSP with research links tend to be further away from the relevant university departments, although when the CSP was set up, close proximity was viewed as an important factor.

The value of links with the university can be assessed to some degree by considering their nature. In the survey companies were asked what main types of local research links they had. Table 6.6 shows the results. The most important type for firms in the CSP was the exchange of ideas, although 13 per cent of the companies engaged in R&D also shared equipment. Although firms in the sub-region also pointed to the importance of exchanging ideas there were two significant differences between the two groups of firms — namely the much greater proportion of firms in the sub-region claiming to share information and equipment and the extent to which firms in the CSP exchanged people.

Table 6.7 shows that when all research links are considered, companies in the CSP generally had an orientation towards other organizations, such as government research bodies and other companies, compared with firms elsewhere in the sub-region. This was true for all types of companies.

Table 6.7 Research links with various organizations of companies at the Cambridge Science Park

	CSP			*Cambridge sub-region*		
Organizations	*All*	*High-tech*	*R&D*	*All*	*High-tech*	*R&D*
Cambridge University	24	23	37	19	32	33
Other educational bodies	8	9	12	8	13	13
Government research bodies	24	27	37	16	26	29
Companies	32	27	50	28	40	48
Miscellaneous	8	14	12	3	6	6

Source: Moore and Spires (1983).

Figures are percentage rates of incidence.

What stands out from the results is the very limited linkages between the CSP tenants and the university. This finding is in line with other work on science parks in the United States. A number of factors are likely to be at work constraining the extent to which firms in the CSP have developed research links. First, several companies on the park are mainly concerned with marketing or distribution rather than with manufacturing or R&D. Thus over a third of the companies located at the CSP do not undertake R&D on site (this proportion is about the same as that for other firms in the Cambridge sub-region). When asked the reasons for location at the CSP, companies did not rank very highly the desire to establish research links. The choice of factors is shown in Table 6.8.

The attractiveness of the area as a place for employees to live in was particularly important, while the availability of labour and premises emerged necessarily as key factors, together with the good reputation of the area. Hence it was not surprising that R&D links have been slow to materialize. The main reason for choosing the science park within the sub-region were the attraction of the image,

Table 6.8 Reasons for locating in the Cambridge sub-region

	CSP			*Cambridge sub-region*		
Individual factors	*All*	*High-tech*	*R&D*	*All*	*High-tech*	*R&D*
Attractiveness of the area for employees to live in	64.0	76.2	86.7	36.5	48.8	47.6
Availability of labour	64.0	66.7	73.3	37.8	41.5	35.7
Availability of premises	60.0	66.7	73.3	44.6	36.6	45.2
Good reputation of the area	56.0	66.7	60.0	36.5	46.3	47.6
Good communications	40.0	47.6	46.7	35.1	29.3	35.7
To establish research links	28.0	33.3	40.0	14.9	17.1	23.8
To be near similar organizations	28.0	28.6	26.7	27.0	39.0	28.6
Robust local economy	24.0	23.8	26.7	17.6	12.2	14.3
To be near customers	20.0	19.0	26.7	29.7	24.4	21.4
Group rationalization	16.0	9.5	13.3	1.4	—	—
To be near suppliers	12.0	14.3	6.7	13.5	2.4	9.5
Government assistance	—	—	—	1.4	2.4	2.4

Source: Moore and Spires (1983).

Figures indicate the importance of individual factors (percentage of companies that chose each factor as either very important or important).

the capacity of the sites and units to accommodate change, and the pleasant working environment. The relationship between these factors is shown in Table 6.9. This emphasis was reinforced by discussions with companies who suggested that a science park address in Cambridge created a good image for the company, and the attractiveness of the area appealed to management/professional and scientific/technical personnel.

For the company sample in the sub-region, a similar ranking emerged in terms of reasons for locating in the sub-region, although the 'R&D links factor' appears to be marginally less important than was the case for CSP companies. The attractiveness of the area is ranked highly again for high-tech and R&D firms, perhaps because of the perceived need to attract managerial/professional and scientific/technical staff.

There are, of course, other reasons why university/industry collaboration may be limited. Physical proximity cannot overcome the problem of mismatch between what the local university can provide by way of help and what the company needs. In a closely integrated country like Britain geographical barriers to effective communication can probably be overcome, as shown in Table 6.10, which gives the regional research links for companies of the science park.

Table 6.9 Reasons for location at the Cambridge Science Park as opposed to elsewhere in the sub-region

	Cambridge Science Park		
Individual factors	*All*	*High-tech*	*R&D*
The park is well managed	52.4	61.1	46.2
Upkeep of the site	38.1	44.3	46.2
Selection of tenants	38.1	44.3	46.2
Servicing arrangements	14.3	16.7	8.0
Attraction of the science park image	71.4	77.8	76.7
Relationship with the university	33.3	38.9	46.2
Design of the site	28.6	33.3	23.1
Design of the buildings	23.8	27.8	15.4
Attraction of the tenancy arrangements	33.3	33.3	38.5
Rental levels	14.3	16.7	15.4
Leasing arrangements	28.6	27.8	30.8
Capacity to accommodate change	71.4	66.7	61.5
Internal flexibility	38.1	33.3	46.2
Land for expansion	19.0	16.7	8.0
Few placing constraints on expansion	9.5	11.1	—
Pleasant working environment	71.4	77.8	69.2
Layout of the park	42.9	50.0	46.2
Design of units	19.0	22.2	15.4

Source: Moore and Spires (1983).

Figures reflect the importance of individual factors, showing the percentage of companies that chose each factor as either very important or important.

Table 6.10 Geographical distribution of research links of companies at the Cambridge Science Park

	Cambridge Science Park		
Region	*All*	*High-tech*	*R&D*
Sub-region	36	36	56
East Anglia	4	5	6
South East	20	18	31
Other regions	24	23	37
Abroad	8	9	12

Source: Moore and Spires (1983).

Figures are given as percentages.

If collaboration is seen to be mutually beneficial it is likely to occur whether 5 or 50 miles separate the organizations.

Another problem is that companies and universities frequently undertake research for quite different purposes. At times the different aims may be contrary. For example, companies are concerned about commercial secrecy at key stages when new products or processes are being developed, whereas university departments usually have an obligation to publish the results of research after a relatively short period even when it has been undertaken on a contract basis for companies. The differing aims can affect the nature of research links, although this depends on the participants and the type of research.

In some ways the orientation of research in university departments is more related to the pure research aspects of innovation rather than development or application. By contrast, companies concentrate on development, especially with regard to product improvement. These two areas are linked and overlap in terms of the mechanics of innovation; however, as different organizations have different orientations, this can reduce the desire to forge links.

There appears to be a general lack of information about the activities and facilities of all organizations involved in research in the Cambridge area. For example this relates to current research activities, the availability of equipment, and specialized information sources. This shortfall of general information probably inhibits the development of links. It is apparent that for many individuals and organizations there are no major incentives to establish research links. These cover reputational and financial aspects such as remuneration and grants. Without firm incentives the possibility of links being established must be more limited. In recognition of the constraints two local initiatives are likely to be implemented in order to improve the existing situation. Trinity College is to build a social centre with facilities for seminars at the CSP which should provide more opprtunities for both formal and informal liaison, particularly between CSP tenants. Equally, they have established a fund to pay up to half the costs for two to three researchers who will spend about half their time with their employers at the CSP and the remainder in university laboratories working on agreed projects. Cambridge University is likely to improve the general liaison with all companies by reorganizing the Wolfson Cambridge Industrial Unit.

Employment opportunities

An increasingly important objective associated with the CSP and the establishment of science parks generally is the improvement of job opportunities in an area. Indeed, many of the recent initiatives in Britain have put far more emphasis on the objective than on the role of the science park as a vehicle for forging research links between universities and industry. Moreover, the objective goes beyond the creation of jobs per se to the important issue of enriching the skill content of the local labour force. In this way science parks, by improving the local labour market environment, enhance the attractiveness of the area for actual and potentially mobile projects requiring technical and scientific labour.

The CSP has undoubtedly provided important opportunities for local firms wishing to start or expand; 44 per cent of the companies are of local origin (with several companies founded by university staff). About 60 per cent of the CSP tenants are linked to overseas multinationals, although only a quarter (accounting for about one-fifth of employment) can be said to have originated overseas. The remainder, about a quarter, come from other parts of Britain. This picture of the 'source' of employment is matched very closely by the companies surveyed in the Cambridge sub-area, where about 50 per cent were spawned locally.

Table 6.11 Occupational structure of companies in the Cambridge Science Park

	Cambridge Science Park			*Cambridge sub-region*		
Occupational group	*All*	*High-tech*	*R&D*	*All*	*High-tech*	*R&D*
Managerial/professional	23	21	20	15	15	15
Scientific/technical	37	38	41	27	34	37
Clerical	19	19	18	23	29	24
Total white collar	79	78	79	65	78	76
Skilled manual	6	6	5	18	16	15
Semi-skilled manual	9	10	9	14	5	8
Unskilled manual	6	6	7	3	1	1
Total blue collar	21	22	21	35	22	24

Source: Moore and Spires (1983).
Figures are given as percentages.

Table 6.11 compares the occupational structure of employment in companies in the CSP and companies in the sub-region. Over 60 per cent of the CSP employees were managerial/professional and scientific/technical staff compared with just over 40 per cent in these occupations for companies in the sub-region. For white-collar occupations as a whole the proportions are 79 per cent and 65 per cent respectively. There is evidence, therefore, of a small but significant bias in the science park of recruitment towards the more skilled occupational categories. The proportion of blue-collar jobs has not, therefore, reached the 33 per cent anticipated when the CSP was set up.

An analysis of the recruitment behaviour of firms as shown in Table 6.12 gives some indication of the extent to which local people are able to take up the employment opportunities offered. In this context local people are seen as those who normally reside within the sub-region. When all firms are considered 45 per cent of CSP jobs were filled by local people and 69 per cent in the remainder of the sub-region. This pattern was repeated by high-tech and R&D companies in lower proportions: 61 per cent and 59 per cent respectively, in the sub-region.

Table 6.12 Recruitment of employees from within the sub-region to the Cambridge Science Park

	Cambridge Science Park			*Cambridge sub-region*		
Occupational group	*All*	*High-tech*	*R&D*	*All*	*High-tech*	*R&D*
Managerial/professional	24	26	26	62	62	58
Scientific/technical	28	28	26	43	38	43
Clerical	75	78	84	71	66	65
Skilled manual	59	60	59	97	90	77
Semi-skilled manual	64	66	69	88	98	87
Unskilled manual	85	85	94	80	100	69
Total	45	46	46	69	61	59

Source: Moore and Spires (1983).

Figures are given as percentages.

This analysis by itself cannot tell us how many jobs would have been lost to the Cambridge area if the CSP had not been established. Certainly the evidence for the sub-region (based as it is on companies establishing after 1960) does not suggest that site availability was a major constraint either on new company formation or inward movement of companies with origins elsewhere, except where these companies were not small initially. It would be fair to say that even though the policy of the local authority has been one of restraining economic development and selective relaxations on the type of development permitted, nevertheless almost 50 companies moved into the area and survived in this period from the sample alone. Moreover, given the similarity and nature of companies inside and outside the science park with regard to their size and activities, and the availability of land and premises elsewhere (even with constraints), together with their reasons for wishing to locate in the sub-region, it is likely that they could have been accommodated in the Cambridge area, although any new development would have been piecemeal.

If there are strong arguments for believing that the development would otherwise have occurred, albeit in a piecemeal fashion, the question of the merits of the science park as a vehicle for job creation rests on the indirect employment effects and the impact on the quality of the local labour force. With respect to the latter it has been shown that there was a tendency for science park firms to recruit a higher proportion of more skilled labour but not substantially so.

The *indirect effects* on employment creation are numerous and difficult to quantify. First there are the conventional local input/output linkages. Table 6.13 provides some evidence on this. A greater proportion of firms in the Cambridge sub-region have local suppliers (47 per cent compared with 36 per cent of the CSP companies); however, there is a slight tendency for firms engaged in R&D at the CSP to have stronger local connections than similar firms in the sub-region. What also stands out clearly is the much greater dependence of CSP companies on imports from abroad compared with sub-region companies.

Table 6.13 Geographical distribution of main suppliers to the companies in the Cambridge Science Park

	Cambridge Science Park			*Cambridge sub-region*		
Region	*All*	*High-tech*	*R&D*	*All*	*High-tech*	*R&D*
Sub-region	36	41	50	47	40	44
East Anglia	16	14	12	27	17	31
South-East	44	41	50	70	55	71
Other regions	36	41	31	36	40	37
Abroad	64	64	62	30	38	31

Source: Moore and Spires (1983).

Figures are given as percentages.

The picture on local linkages is further complicated by the fact that although fewer CSP companies had local suppliers, they were on average larger companies on the basis of turnover, as shown in Table 6.14.

Local linkages also appear to be greater for firms in the sub-region in terms of main customers. Table 6.15 shows that 47 per cent of companies in the sub-region had main customers locally, compared with just 16 per cent of firms in the CSP. However, the

Table 6.14 Value of sales of companies with main suppliers in the Cambridge sub-region

£ millions	*Cambridge Science Park*	*Cambridge sub-region*
<0.5	50	74
0.5–1	20	7
1–2	—	10
2–4	10	—
>4	20	10

Source: Moore and Spires (1983).
Figures are given as percentages.

Table 6.15 Geographical distribution of main customers

	Cambridge Science Park			*Cambridge sub-region*		
Region	*All*	*High-tech*	*R&D*	*All*	*High-tech*	*R&D*
Sub-region	16	14	12	47	40	33
East Anglia	8	9	6	30	28	21
South-East	56	55	50	54	57	62
Other regions	72	73	69	57	60	67
Abroad	64	73	75	48	64	58

Source: Moore and Spires (1983).
Figures are given as percentages.

Table 6.16 Value of sales of companies with main customers in the sub-region

£ millions	*Cambridge Science Park*	*Cambridge sub-region*
<0.5	25	76
0.5–1	25	5
1–2	50	10
2–4	—	2
4	—	7

Source: Moore and Spires (1983).
Figures are given as percentages.

average turnover of the CSP companies is again somewhat larger than that of the sub-regional companies. Table 6.16 gives the details.

A second source of indirect employment effects relates to the growth of population brought about by industrial development. The analysis above shows that a significantly greater proportion of CSP firms (than sub-regional firms generally) recruit employees from outside the sub-region. Insofar as this creates a change of residence rather than merely a change in the pattern of the work, the CSP will have resulted in a differentially faster growth of the local population compared with companies in the sub-region. Increased population means a higher aggregate local income, and although there will be substantial leakages of expenditure out of the local area notable expenditure on services will create local income and employment.[1]

Another factor to consider when comparing the indirect employment effects of a CSP-type industrial development with development in the sub-region is the effect of the higher average incomes of a typical science park employee, because of the occupational bias of the labour force. The more skilled groups operate in a national labour market and as such local firms will be obliged to consider national as opposed to local pay levels as their guidelines when recruiting labour. To some extent, of course, the attractiveness of the area will reinforce or offset these tendencies, and certainly in the specific case of the CSP this has been important (see Table 6.11).

It is clear that a variety of employment effects are likely to arise. It is difficult to assess the differences between employment effects of the CSP and those arising from industrial development in the sub-region more generally. Some factors such as the occupational structure of the CSP tend to reduce relative local employment impacts, whereas others such as induced population growth tend to raise them. At this stage it is not possible to assess the balance of these effects.

However, the CSP has been a high consumer of industrial land because of the desire to create a very low-density development within well-landscaped surroundings. The effect has been that the CSP has utilized almost twice as much land as conventional industrial estates for the same amount of floor space. The typical ratio of floor space to land in the CSP is approximately 1:6. On the other industrial estates it is generally 1:3.

Consequently, the land allocated for the CSP development could have been utilized to generate twice as many jobs if a more conventional form of development had been permitted, based on the same ratio of employees to floor space. The foregoing analysis suggests that a higher proportion of these jobs would have been taken by people in the sub-regions. Overall the land may have been capable of providing almost 1,380 jobs for local people compared with 445 provided by CSP firms.

Conclusion

The above analysis of the experience of companies in the CSP has provided some valuable initial insights into the effectiveness of this kind of initiative in certain key respects. First, it is clear that the science park did succeed in acquiring tenants whose activities are heavily biased towards R&D high-technology. However, the comparison of CSP companies and a random sample of companies established elsewhere in the sub-region over broadly the same period of time reveals that the two groups are very similar. In terms of industrial structure, size, occupational structure, and reasons for choice of location in the sub-region they are similar. However, there was a tendency for the companies opening in the sub-region to recruit a greater proportion of their labour from the local labour market.

Second, two key objectives for the CSP were identified — the promotion of greater research links between industry and the university and the creation of improved employment opportunities in the local area. To some extent both of these objectives were secured. About one-quarter of the companies in the science park had developed links of one kind or another with the university. However, the links forged are no more prevalent than for firms outside the park, although there is some evidence that more links have been established for firms engaged in R&D. The survey also shows that very close proximity is not a necessary condition for establishing research links between industry and the university.

With respect to employment creation, it is clear that the occupational structure of the firms attracted to the science park is biased towards white-collar workers and, in particular, the professional and technical groups. In the absence of the CSP there is a strong presumption that a similar rate of job creation would have been secured albeit in a piecemeal fashion. Because a greater proportion of recruitment by firms in the CSP relative to firms in the sub-region took place outside the sub-region, their direct employment effects on the local labour market are correspondingly less. Typically, CSP firms recruited about a quarter fewer jobs locally than firms opening in the rest of the sub-region. However, indirect effects partly offset this, in that the population growth associated with CSP recruitment patterns was faster and earnings of employees in CSP firms are also somewhat higher. Both of these factors increase employment relative to the outcome for firms in the sub-region. Of much greater potential significance, however, are the employment consequences of very low density industrial development in the science park. Conventional industrial development would have produced three times as many jobs for local people compared with the CSP. Evidence from the companies in the sub-region suggests that such low-density development is unnecessary, although it is undoubtedly an attractive feature of the science park.

Heriot-Watt University Research Park

Summary

Heriot-Watt University established its science park in 1972 with the prime objective of encouraging the transfer of academic research into marketable industrial processes, products, and services. In particular, the research park seeks to encourage the specific activities of research, development, design, engineering, and prototyping by individual company staffs in close proximity to the facilities and the academic community of the university. The park, which is financed only by the university, is an integral part of the campus of Heriot-Watt and provides particularly favourable opportunities for close collaboration between companies and university departments.

This case study examines the objectives of the park; its role within the wider scope of collaboration with industry; its resources and facilities; the source and nature of the companies that have located there; and university encouragement of academic members of staff to form 'spin-off' companies. Finally, the case study presents conclusions, comments, and a review of future plans.

Introduction: Heriot-Watt University

Heriot-Watt University is located in Edinburgh, Scotland, and its origins can be traced back through a century and a half of achievement in technical education and effective co-operation with industry in training and research. In the evolution of the university, two principal characteristics emerge: the maintenance of rigorous academic standards, and the relevance of all training and research to the needs of industry and the community. These are the foundations on which the university has established its present reputation as a leading institution in the development of science and technology in Britain, and on which it is continuing to build.

With this philosophy it was natural for the university to encourage members of the academic staff to maintain close collaboration with industry, and these links have been built up steadily throughout its history as an engineering and science institution. This approach was further developed by the decision to form dedicated technology transfer groups in particular subject areas. In

1968 the university established the Computer Application Services, an applied research group that has been notably and continuously successful in the field of information technology. In 1972 the Institute of Offshore Engineering was formed, and other groups have followed in such diverse fields as economics, distance learning, and medical laser technology.

In 1966, in order to facilitate expansion and future development, the decision was taken to remove the whole university to a new green field site of 90 hectares at Riccarton. The first of the academic departments moved to the new campus early in 1973, and it is expected that the process of removing the whole of the university from the city centre to the new site will take until 1991.

The availability of its new Riccarton campus enabled the university to establish its research park in 1972.

Development of a policy for collaboration between Heriot-Watt University and industry

The planning of collaboration between the university and industry was based on the desire to take a firm initiative in this matter, suited to the requirements of the community in which the university is located.

The pleasing environment of Edinburgh and the Lothian region has been a powerful attraction to new industrial ventures and in particular has given encouragement to the clean industries, such as electronics, health care, and precision engineering. More recently the support of exploration and exploitation of the oil and gas reserves in the Scottish area of the North Sea has introduced a further range of industries concerned with offshore technology. As a result, the last 30 years have seen a remarkable growth of such new, science-based companies in this area. These comprise both foreign (predominantly US and Japanese) and domestic investment, and include a significant number of local enterprises which have developed as a consequence of the presence of the large companies.

As a modern engineering and science-based institution, the university has a role to play in all of these developments. Undergraduate and postgraduate teaching, together with pure research, is clearly the major contribution to be made, and this will continue to be the prime concern of the university. Heriot-Watt wished to extend this activity, however, by providing more immediate support to industry as a means of improving its capabilities by introducing new methods as rapidly and directly as possible. These new techniques would be founded on the knowledge and skill of the staff of the university, and the employment of its resources in the continuous transfer of technology.

Thus the philosophy was developed of creating a bridging activity to transfer new information and methods from the university into established practice in industry. This policy has taken three major forms:

1. Active encouragement, by the university, of departments and individual members of academic staff, to make available their services and resources to industry on a professional basis. This has been aided by the establishment in 1969 of an industrial liaison unit, called Unilink, to the main centre for fostering and strengthening links between the academic departments of the university and industry.

2. The creation, by the university, of dedicated services in specific activities in which the demand requires that staff be recruited solely for this purpose in order to be able to devote all of their time to these tasks. Such activities (e.g. Computer Application Services, Institute of Offshore Engineering, Marine Science Unit, Economic Research Unit, Medical Laser Unit, etc.) are an integral part of the university but are concerned with providing an outward educational R&D service to industry.

3. The creation of a research park as a part of the university's new campus. It is important to appreciate that all three forms of activity are complementary and reflect the different degrees of collaboration that may be appropriate to any one situation. This coherent approach is important since it extends the intentions behind the formation of a specific area of the university campus as a research park into a very much wider area.

The research park

The creation of the research park was the initiative of the university alone. In an area of predominantly science-based industry and ample modern industrial estates, it was not considered appropriate for Heriot-Watt to promote a high-technology production site of the type usually referred to as a science park. Instead, it was decided that applied R&D was the most constructive sector in which the university could make a contribution to active collaboration with industry. It was realized, however, that production will often arise naturally from the result of R&D activities, and it was decided that 'the limited manufacture of sophisticated items containing a large element of development, not for sale to the general market, be allowed'. In addition the university requires to be satisfied that the activity envisaged is of a type that will promote interaction between the tenant and the university, particularly in R&D.

The university initially allocated some 9 hectares (approximately 10 per cent of the area of the campus) for this purpose immediately inside the main entrance to the campus. This prominent position for the research park emphasizes the university's commitment to its policy of active collaboration with industry, and ensures that the companies occupying sites in the park clearly see themselves as part of the university campus rather than a mere adjunct to it. The first tenant of the Heriot-Watt Research Park was located on site in 1972.

The research park offers a number of advantages both to the companies that place their laboratories on that site, and to the university. These may be listed briefly as follows:

The research park offers the most suitable environment in which to carry out research and development activities.

The location of the university and the advantages of the surrounding area are such as to prove attractive to staff and their families as a place to live and work, thus making it easier for companies to attract the most talented staff.

Frequent opportunities occur for informal contacts to be made and maintained between company staff and academics from relevant disciplines (in the restaurants or sports facilities for example), thus enabling an easy and frequent exchange of non-confidential scientific and industrial information to take place to the benefit of both parties.

The companies have ready access to university facilities such as libraries and special equipment which would be extremely costly or economically unjustifiable on an individual company basis.

The convenient opportunity for companies to make use of any of the services offered by the university (including the library, computers, and specialized equipment); to retain on a formal basis appropriate members of the academic staff; and to sponsor specific items of research and postgraduate studies.

The presence of the companies and their staffs widens and enriches the interests and experience of the academic staff, and through these contacts the latter are enabled to maintain a closer understanding of the needs and current areas of interest in the appropriate industrial sectors.

> The presence of the companies adds both relevance and opportunity to the teaching and work of the undergraduates, and stimulates research in topics of mutual interest.
>
> The park furnishes both the means and the encouragement for the further development of ideas and inventions by members of the university staff through the setting up of 'spin-off' companies.

It is a matter of satisfaction to all concerned that the above advantages foreseen in the creation of the park are being realized in practice both by the companies with laboratories located on site, and by the university.

The financial investment in the construction and servicing of the research park is a large one and this has been made from the university's private resources. There is also a significant and continuing expense involved in the development of the park, the location of companies, and negotiation with potential occupants of the park. The sum of these costs to date is of the order of £500,000 (in actual terms) which value would have to be increased by, say, a factor of 3.5 to give a current figure. Such a project must be seen, therefore, as a long-term development, requiring a continuing and enthusiastic commitment if its potential is to be fully realized.

At the present time, some 90 per cent of the original area of the park is allocated, including options on sites for further development. A matter of concern to every company is adequate provision for future expansion and at present, with only one exception, all of the organizations located on the park have either completed, or are in the process of building, the second phase of their development.

At present, there are 18 campus companies located at Heriot-Watt, and active discussions and negotiations are in hand with several more. In total, they represent research, development, and engineering activities originating from three main sources in more or less equal proportions, namely university technology transfer groups, 'spin-off' companies from university departments, and external companies. It is the intention of the university to preserve this approach and this composition of park activities in the future. Three major sectors of industrial interest have emerged among the park occupants, and these are health care, offshore/petroleum engineering, and computer/electronics/electro-optic engineering.

Further progress has been made through an agreement with the Scottish Development Agency whereby it has financed the construction of a multi-occupancy laboratory building. These units share common facilities such as reception office, conference rooms, and so on, within a building that is designed to be subdivided readily to meet the needs of a number of occupants. This facility has proved to be an attractive one, and negotiations were very quickly completed with three organizations regarding the combined letting of all of this laboratory space. The university plans to provide further laboratory units of this type.

As evidence of the confidence that the university has in the future of the research park, the opportunity was taken five years ago to acquire a further area of some 9 hectares, contiguous with the present site, thus doubling the total area available for park development. It is foreseen that it will be necessary to open up this second area for development in the near future, and the provision of the necessary infrastructure and services will require a further significant capital investment in the park. Indeed, it is reasonable to expect that investment in these latter items will have doubled (at present values) by 1990.

Factors concerning the development of the research park

As a pioneer of the concept of science parks in Britain, Heriot-Watt undoubtedly experienced a prolonged initial period in which the opportunities offered by such a facility were slow to be appreciated by industry and the community at large. In the last five years or so this attitude has changed radically, and there is now a much greater awareness and understanding of park objectives. Thus, the growth of the Heriot-Watt Research Park, if measured in numbers of companies, has been exponential. Persistence, and a constancy of purpose, are clearly necessary, and it has always been accepted by the university that the development of the park would be a long-term project. In the case of Heriot-Watt, growth has also been affected by the unavoidable circumstances of twin-site operation (a separation of some 10 km) while the whole university is re-located at the new campus. This move is planned for completion by 1991, and the resultant concentration will bring with it the related collaboration not presently sited at the Riccarton campus. Indeed, if present rates of development are maintained, then phases 1 and 2 of the park will be totally assigned by the mid-1990s. By that stage, recovery of the initial capital investment, through the mechanism of ground rents paid by tenant companies, will proceed at a relatively rapid rate.

It must be emphasized, however, that the purpose behind the formation of the park is not financial, and that the measure that the university applies to this project is that of the degree of interaction that takes place between tenant companies and the staff of the university. Here it should be noted that it is the individuals who matter, and since these change from time to time through promotion, postings to other parts of the company, or a move to a new employer, there can be no such thing as a once-and-for-all corporate introduction made between a tenant company and university staff. The process of widening the range of acquaintanceships and the introduction of new members of staff from both sides is a continuous task. And it is natural to assume that the focus of collaboration will be on the prime discipline of the tenant company. Although this is more often the case with the smaller companies in the park, the lesson of the years has been that, more often than not, it is a range of subsidiary expertise which forms the most fruitful area for co-operation. It is not unknown for one company to have as many as five channels of collaboration in operation simultaneously in subjects other than its prime disciplines. Measured in these terms, the park has been a success from its earliest days, and there is no reason to doubt that this will continue to be so in the future.

The category of park occupant in which interaction with university departments is most readily achieved is that of the 'spin-off' company, of course. This term is used to describe companies that have been formed by, and are owned either in total or in part, by members of the academic staff in order to carry through the results of their academic research into marketable products and processes. Such companies are financially and legally separate entities from the university, but every encouragement is given to promising projects of this type. Clearly this is a most desirable and direct method of 'shaking out' good ideas from academic research and using them to form the basis of new activities. In order to achieve this, however, it is necessary for the academic member of staff to adjust psychologically to the different nature of employment, sources of income, and additional considerations (such as financing) that will affect his or her working life. One mutually satisfactory arrangement, which has been successfully used by the university, has been to agree to a part-time employment contract with the academic entrepreneur whereby the latter has a basic income ensured, plus the other half of his time to devote to his company. In return, the university still has access to his teaching skills, together with the use of the half-salary to employ 'new blood' teaching and research resources. Needless to say, the opportunities for collaboration in such circumstances are maximized to the benefit of all concerned. Examples of such activities range from tunable spin-slip laser technology to heat pumps.

It is typically the case with new start-up companies that there is a need for wider management advice in the setting-up of the company, in compiling business plans, in locating suitable sources of finance at the various stages of the company's development, and in presenting such proposals to potential investors. The park directorship provides such assistance when and where it is required, aided by the fact that the city of Edinburgh is a major financial centre with ample venture capital funds.

Such support, advice, and information can significantly strengthen and augment the usually inherently high scientific and technical capabilities of the new spin-off company. There remains, however, the problem common to all new, small companies, namely that of financing a marketing effort strong enough to ensure that the products that have been developed can be sold in sufficient quantities, and at a sufficient rate, to yield an adequate and timely return. This problem is exacerbated in the case of research park tenants in that typically the market for high technology is an international one with consequently high marketing costs. There is no simple answer to this problem, but at least the sources of advice referred to above can ensure that the problem is appreciated in advance, that adequate provision is made to deal with this activity, and that the attention of management can be focused on this task without the danger of other unconsidered aspects of the company's operations compounding the difficulty. It is also worth noting that, without intruding upon the confidential information of any one company's activities, peer occupants of the park can find assistance and encouragement in shared experience of this (and other) problems, and thus the park community's progress along the 'learning curve' is improved to the benefit of all.

The latter comment leads naturally to the interaction that can take place between tenants of the research park. Clearly there are occasions when equipment belonging to one company may be made available to another, but more significantly there are occasions when the products of one can be of value to another. While it would be wrong to suggest that this happens frequently, neither is it true to say that such cases are rare. The example may be quoted of one tenant successfully developing a novel production process for a material and, as a consequence, opening up a prime opportunity for another tenant company to develop some of the specialized equipment that is required by licensees of the process for the manufacture of the product.

Furthermore, examples have emerged of 'spin-off' company development of an academic research topic leading to the design and subsequent production of an instrument product of advanced performance. The academic research group has then acquired one of these instruments, and has been enabled to make further advances in its field due, in part, to the availability of the developed and engineered instrument. Such mutually advantageous and supportive interaction can yield rapid and significant progress for both parties. Although these may represent particularly striking examples of mutual interest and interaction between tenants within the park, they do underline the lesson that the opportunities for collaboration do exist and only require the right attitude of mind to ensure that benefit is obtained from these opportunities.

Finally, it might be interesting to note that no company located on the park to date has failed and gone out of business. Leaving aside for the moment any question of whether or not the advantages of a location on the research park have been the deciding factor in this happy result, the encouraging effect of this atmosphere of success serves to stimulate the spirit of emulation in others and thus helps 'new things to happen'.

Conclusions and future plans

The university intends to develop the research park in keeping with the general ideas outlined earlier, and now summarized.

The fundamental activities of the university are the pursuit of learning, teaching, and research. In this way, it makes a major contribution to the future of the country as a whole, and to industry in particular.

As a science-based institution it is vital that the university makes certain that in providing the trained minds and the knowledge on which the future will be built, there is a true understanding of current needs.

The university believes that not only does this approach ensure that both the academic staff, and through them the undergraduates, maintain an up-to-date understanding of the needs of industry but that industry is well served in the process.

By maintaining and developing its contacts with industry and other bodies; by encouraging individual members of staff to undertake consultancy work that is related and complementary to their fields of interest; by initiating new research projects and courses of direct relevance to the needs of industry and the community; by establishing technology transfer groups; by encouraging the formation of 'spin-off' companies; and by offering the special environment of the research park to external companies, the university continues to be concerned, not simply with the pursuit of academic knowledge, but with fulfilling its role as a technological institution in the wider community that it exists to serve.

Finally, the university intends to continue the expansion of the research park, and to encourage new developments, while keeping the essential character of the Riccarton campus. In so doing, it will continue to encourage the composition of the park to be a mixture of external, 'spin-off', and technology groups working in close collaboration with the university.

The Alvey Programme

Actual expenditure (£ millions)

1984–85	1983–84	1982–83
6.45	1.73	0

The Alvey Programme is a five-year programme of collaborative research in the enabling technologies of information technology (IT). Projects are jointly sponsored by the Department of Trade and Industry (DTI), the Ministry of Defence (MOD), and the Science and Engineering Research Council (SERC) (under the Department of Education and Science) together with the industry. It has an overall budget of £350 million, of which more than half will be provided by the government. Its objective is to ensure that Britain maintains an internationally competitive position in advanced information technology towards the end of the century. The programme should double the level of IT research in Britain over its five-year-life.

Origins of the programme

The key feature of the programme is collaboration at project level between companies and academic institutions, and at management level between government departments and between government and industry. Projects must normally involve participants from at least two companies together with academic institutions as appropriate. The programme is administered by a team drawn from DTI, MOD, SERC, and industry.

The programme is a response to increased overseas effort on information technology research, particularly in Japan, which announced its programme in April 1982. The government decided to appoint a committee chaired by Mr John Alvey of British Telecom to advise on the scope for a British programme. The Alvey Committee drew attention to the dangers of an increasingly adverse balance of trade in IT products, and warned that unless the British

IT industry maintains a strong competitive position the efficiency of other industries in manufacturing and services would suffer. It pointed out that IT research is increasingly expensive and that fewer companies can afford it without co-operation, which helps reduce barriers to high-risk, high-cost research.

Technical coverage

The programme is one of directed research. It covers four main areas, for each of which there is a detailed technical strategy against which individual projects are matched.

The first area is intelligent knowledge-based systems (IKBS), the term adopted for what is often called 'artificial intelligence'. The objective is to stimulate IKBS research in Britain and the market for its results. The research themes cover architectures for parallel processing, logic programming, knowledge bases, and expert systems. In addition there is a programme of software development to support the necessary specialist languages such as Prolog and LISP. Complementary awareness activities concentrate on the formation of expert-system clubs for information exchange and to help spread expert techniques in such fields as process control, quality surveying, and banking.

In the next area, software engineering, the objective is to establish tools and methods necessary for the production of high-quality cost-effective software and to ensure its widespread use. The main approach is through the establishment of successive generations of integrated project support environments (IPSE), and formal methods for programme specification and design.

In very large scale integration (VLSI) the objectives are to achieve internationally competitive VLSI processes and a national approach to computer-aided design for VLSI. Targets have been set for greater miniaturization to allow higher densities of components on the chip and higher speed of operation. Computer-aided design (CAD) is crucial to the exploitation of VLSI circuits, and effort is being put into developing the required CAD tools.

Finally, in the area of man/machine interface (MMI) the objective is to improve the interaction between machines and their users, in particular to develop techniques of speech and image processing. This is vital to the competitiveness of products. Several projects form important links with IKBS since, for instance, a high degree of machine intelligence is required to understand speech or interpret images.

The programme also supports large-scale demonstrator projects aimed at stimulating enabling technologies into creating practical application, and to show in visible form the practical exploitable results of the programme. There are currently four projects, each led by a major industrial company. One, called 'Design to Product', is led by General Electric Company plc and will demonstrate the automation of the total production process, in this case of a fuel pump, from design through manufacture to maintenance in the field, using computer-integrated manufacturing and IKBS techniques.

International Computers is leading another demonstrator in the Department of Health and Social Security (DHSS). This is studying the role of new computing technologies, such as expert systems, in helping legislation-based organizations to formulate, interpret, and apply complex policies and rules.

Plessey leads the Speech Input Wordprocessor and Work Station project. This aims at practical recognition of connected speech with a vocabulary of 5,000 words at minimal error rates. Products derived from it could include wordprocessors for office use among many others.

A Mobile Information Systems project is led by Racal. This involves the design of a range of highly intelligent mobile terminals that can communicate with various public and private databases via radio links.

Finally, the programme supports advanced work on parallel processing architectures and data communications. An Advanced Network Systems Architecture project has been approved to establish a system architecture that allows fifth-generation sub-systems to interact efficiently and easily from the user's point of view. The project is centrally located and involves several companies.

Programme management

The Alvey Directorate, based in London, manages the programme and appraises individual projects. In addition to the programme Director there are seven other directors, each responsible for an aspect of the programme including administration. The directorate comprises officials from three government departments involved in roughly equal proportions, together with industrial secondees. Their work is overseen by a small steering committee of industrialists and government representatives chaired by Sir Robert Telford (formerly chairman of Marconi). The government funds are provided by SERC (£50 million), MOD (£40 million) and DTI (£110 million); an in-house publication, the *Alvey News*, is published every two months and is available to all the organizations involved in Alvey projects, together with any others interested in keeping up with programme developments.

Eligibility and conditions

Any centre of IT research in Britain of any size, public or private — whether industry, academic institution, or government research establishment — can participate. There is no bar to the involvement of foreign-owned multinational companies. However, a foreign-controlled company must have a British industrial collaborator; the research work must be done in Britain and the route to exploitation in Britain must be clear. Overall, a consortium will be judged by the balance of advantage to Britain, and British partners will be consulted in this process.

Proposals for projects are judged according to their relevance to the objectives of the programme; the benefit of co-operation they demonstrate; the quality of the participating teams and the background knowledge and experience they brought to the task; and the ability of participants to exploit the work.

The industrial partners receive 50 per cent of the costs of a project. Academic partners are funded 100 per cent through SERC.

Partners in projects are grouped into Alvey 'Clubs', which meet regularly to review progress and provide constructive feedback. A key concern in the programme is that information should be exchanged and research exploited and disseminated as effectively as possible.

The industrial property rights stemming from Alvey projects are owned by the industrial participants, who make appropriate payments to their university collaborators. The arrangements are set out in collaboration agreements that contain the terms for recompensing the academic participants for their contribution. The collaboration agreement is a matter for participants to decide, provided its terms do not inhibit exploitation.

Progress

In each area of research, strategy documents have been published. In addition to the four fields outlined below, the field of computer architecture has been chosen as one requiring particular attention and a strategy has been published. The programme has so far attracted over 550 applicants. Some 100 full collaborative projects have been approved, representing a commitment of about 80 per cent of the programme. The progress of each part of the programme is summarized below.

Very Large Scale Integration

The technology of the VLSI programme is almost entirely committed, except for some funds held for project extension. The Alvey programme, as a co-operative attack, can be seen at its best in the VLSI technology projects, where almost all the relatively few firms that are in the field have combined to co-operate on the enabling technology. This was probably aided by the clubs that predated the Alvey programme by a year or two. This coming together to plan co-operation is a feature that should outlive the programme. The architecture section of the programme, largely signal or sensory data processing, is now fully committed, and the CAD programme is 75 to 80 per cent committed. The remaining funds are earmarked for the second phase of the CAD work, where it has always been envisaged that a single system covering all the main technologies and individual processes would emerge. This is now close to formation.

Software engineering

There are now some 16 fully co-operative projects in software engineering. It has proved to be hard work putting the programme together. This seems to be because there is not the tradition of research in software engineering that exists in other parts of the industry, such as VLSI. The academic work in Britain in this field is of world class, and the Alvey Programme has served to bring this work together with industry. After a slow start it has proved possible to put together a programme that covers much of the strategy drawn up by the community. The work on the two first-generation project support requirements is going well, and a second-generation project is now coming together.

Intelligent Knowledge-Based Systems

Virtually all the IKBS funds, except those set aside for architecture, are now committed. There are some 23 projects, but the average number of parties per project, at between two and three, is lower than for other parts of the programme. The key problem is shortage of experienced staff in industry. The academic community provides high-quality people, but they are limited in number, and the good centres are overloaded. In the case of one part of the strategy, natural language, it has proved possible to generate only a couple of projects with industrial involvement.

Man/machine interface

The man/machine part of the programme is not really a single entity, though the parts are loosely coupled by the concept that the human lies at the heart of the work. Overall, this field is the least developed of the programme, if only because the subjects are most immature. About 60 per cent of the funds are committed.

The speech programme took some time to bring together, but with the help of the staff from the Joint Speech Research Unit, is now quite a respectable if somewhat small programme. The funds are 90 per cent committed in speech, with most of the community involved. The importance of the Speech Club in this field is clearly to pull the 10 projects into a self-supporting community. The large demonstrator concerned with a speech-driven word processor is central to this programme.

The image programme is not quite so satisfactory, since the community is very scattered and its members are not even well acquainted with each other. Nevertheless, quite a healthy programme is coming together, with about 70 per cent of the funds committed. It is noticeable that the industrial community is much thinner than the academic, and the participation of small, specialist firms in several of the projects is interesting. There are two good image projects in IKBS, which depend on an AI-type approach.

The human interface has proved the most difficult part of the Alvey programme to get going. There are a number of projects that have been approved and about half the funds are committed, but the community is very variable and largely consists of inexperienced people, new to the field. However, the Alvey Programme has had an effect and teams are building up in certain firms, albeit rather slowly. When one compares the British situation with that in the United States, where human interface work has now become of very considerable importance in firms from IBM down to small firms with specialist expertise, it is clear that Britain is in danger of slipping behind in this field.

Evaluation

The outstanding achievement of the Alvey Programme has been to increase co-operation throughout the IT community. Typically there are two or three firms in each project, together with one or two universities. Co-operation between firms and academic teams has increased considerably, with universities involved in most of the industrial projects. The work is concentrated in a number of the universities, such as Edinburgh, Cambridge, and Imperial College, though some 50 universities and polytechnics are involved in some part of the programme. Altogether some 60 firms are involved.

Many of the Alvey projects are in their early stages, but already the programme is having a major effect on the way the British information technology community works and plans together. Some of the thinking in the projects is already beginning to carry over into work in the firms aimed at the market, though products directly derived from the programme cannot be expected for some years, except perhaps in the software field, where some of the work on tools may lead to products in a couple of years' time.

The Directorate has set up arrangements to evaluate the programme as it goes along. Three university teams from Sussex, Manchester, and the London Business School are carrying out the evaluation. They are looking at the strategic impact of the programme at the macro-economic level, the impact of the firms, and the administration of the programme itself.

Salford University Industrial Centre Limited

Recognizing the management and structural problems faced by a university in trying to build a relationship with industry, Salford University set up, in 1967, with help from the then Minister of Technology what is now Salford University Industrial Centre (SUIC) Limited.

SUIC is a company limited by guarantee, whose profits are covenanted to Salford University. It is thus legally, organizationally, and managerially quite independent of the university and has to trade at a profit like any other company in a marketplace defined by the Companies Act. It has a full-time general manager; 50 other full-time employees (most of whom are professional engineers); and a board of directors who are part-timers and include the vice-chancellor, registrar, and two professors of the University of Salford, as well as local industrialists, with a part-time non-executive chairman who is currently a local businessman.

SUIC sells technology and consultancy services to industrial, commercial, and other organizations in a wide variety of areas. It has its own building on the university campus, which was opened in 1982. The turnover of SUIC had been growing steadily and is now approaching £4 million. Most of SUIC's activities can be described as putting together and then managing mixed teams of academics and SUIC staff. The SUIC staff know that they have to deliver a specified product, to cost and on time, to their customers and can be sued in the courts if they do not (although they have never actually had that experience). They know where, among their academic colleagues, whether in Salford University or in the other institutions in the region, the needed technical or scientific knowledge resides and they know — too well in some cases — of the difficulties involved in tapping that resource. They largely use that well-known device, money, to get what they want, when they want it, and last quarter over 80 per cent of the individual consultancy earnings of members of staff of Salford University passed through SUIC's books.

A good example of a product arising from a Salford University invention is the piezo-ceramic buzzer. Prototype buzzers made for demonstration at the university were shown to Ward and Goldstone Limited. The invention was sold to the company, which has now established an electronics division manufacturing a range of buzzers, sirens, and alarms for the domestic, commercial, and industrial security markets. Consultancy — to enable companies to improve performance, to develop an opportunity, or to avoid a poor investment — is an equally successful part of the centre's operation. On a recent project for a company in the GEC Group, the centre appraised two new products while they were still at the prototype stage. For each product the consultancy showed how the number of components could be reduced by a fifth and how manufacturing costs could be cut by a quarter. The company estimated savings of production costs of between £50,000 and £100,000 a year.

Professor John Asworth, the university vice chancellor, who took over in 1981, says he has 'made legitimate' the idea that a British university should have transferring technologies to industry as a preoccupation.

Partly to make up for the shortfall caused by spending cuts, Salford has gone all-out to raise money through industrial contracts. Other companies with which it works include Ferranti, Turner and Newall, Rolls-Royce, and Thorn-EMI.

In 1984–85 Salford obtained about 12 per cent of its total income of £25 million from industry, the highest figure among British universities. In contrast the others, which between them cost about £2 billion a year to run, mostly from public funds, earn only a few per cent of their revenues from companies.

Of all the universities in Britain, Salford is probably the closest to industry.

Newtech (Clwyd) Ltd

Following the decimation of traditional North Wales industries (slate, coal, steel production, textiles, brickmaking, agricultural smallholdings) it was realized that progress could only be achieved by ruthlessly sweeping away the debris remaining, physically and mentally, and adopting a radical approach towards building new industries.

Clwyd started its fight back with the conviction that initiatives successful elsewhere in the world cannot be automatically transplanted. On the other hand, Clwyd County Council examined closely what had been achieved: the British science, research, and technology parks; US models from Cambridge and Princeton, the Raleigh Triangle, Silicon Valley, and the new Sun Belt developments; Canadian discovery parks as in British Columbia, to the ultimate technopolis in Japan, Tsukuba Science City, with its 42 research institutes and universities. This investment of the profits of the consumer electronics export bonanza of the 1970s has already attracted industries by the score. A further queue now awaits the use of land released there when Expo-85 closed in 1985.

Compatibility is the major factor in the success of any transplant. The various agencies in North Wales responsible for industrial development readily accepted this fact and consolidated their efforts: Clwyd County Council, the Welsh Development Agency, the Welsh Office Industrial Division, the Deeside Enterprise Trust, and the North East Wales Institute (NEWI). One result is Newtech Square, an attractive complex at the centre of the Deeside Industrial Park costing £2.12 million. This development can be regarded as a very special type or first-phase science park, and in this respect is currently unique in Europe.

The Newtech Innovation Centre forms the corner-stone for development, a one-stop shopping place for innovation, applied research and development, training, technical service, and advice to companies, both large and small. The centre can effectively harness all the resources that individual members of the founding consortium can provide. Additionally, the in-house facilities of the NEWI Research Division, already used extensively by industry, are available for professional expertise and the latest instrumentation in biotechnology, microbiology, and cell biology; analytical chemistry; hydrocolloid, colloid, and surface chemistry; fast reactions, radiation, and photochemistry; food and pharmaceutical chemistry; materials and corrosion science; microelectronics and process and systems control; robotics and computer-aided design and manufacture. As a founder-member of the European Network of Innovation Centres, Newtech can provide ready access to grant aid, technology, and technological markets within the European Community. Newtech's wide range of commercial activities is managed by Newtech (Clwyd) Ltd, established with the specific objective of promoting innovation.

From within the NEWI Research Division an array of industrially sponsored research projects are now achieving commercial application. One example can illustrate the link between NEWI

Research and Newtech. A £250,000 grant from the Biotechnology Directorate of the Science and Engineering Research Council and a subsidiary of the Grand Metropolitan group, Express Dairies, set up a project to develop monoclonal antibody-based immunoassays for use in the food industry.

Based on this work, kits were developed for a rapid, cheap, and revolutionary method of food analysis. Grand Metropolitan group quickly established a new company, Biokits Ltd, to manufacture these innovative kits. This new company, is now located in Newtech Square, where the close link between NEWI Research and the company of developing new assays can continue.

The Newtech Square complex is a highly flexible design, which can house new high-technology companies like Biokits Ltd. Entrepreneurs with innovations benefit from the wide range of services and technology. The surrounding Deeside Industrial Park is already home for 40 new rapidly growing industries. Its location, with special development area status, links the M56 motorway, 35 minutes from Manchester Airport, with post and railway sidings facilities alongside, together with an experienced pool of non-striking labour, make it a prime development site. The Newtech Square complex is now full, with 15 companies either starting up afresh or new developments from existing large companies. They are all R&D orientated.

Newtech Square now provides research, technical, and business facilities not as readily available in any of the traditional science, research, or industrial parks. Such a close integration of academic research and industrial R&D, with sound financial, technical, experimental and business support, in many ways simulates the Tsukuba developments.

In March 1985 Newtech undertook a presentation on its activities to leading Japanese industrialists and researchers. During this follow-up seminar it is with pleasure that a formal link will be finalized between the North East Wales Institute and Gunma University, Japan, part of which will be research in polymeric materials. As a result there has grown up a mutual conviction that Japan can benefit from British basic research, and Wales can acquire some of Japan's development and total quality control expertise.

One of the first joint Japan–Wales projects, with WINTECH support, is a comparative study of education–industry interactions in the two countries. NEWI's links with the University of Wisconsin-Stout will enable the triangle to be completed, since Dr Robert Swanson, chancellor of Stout, currently heads the US Universities' Task Force on the Economic Benefits of University

Research. A seminar comparable to that arranged in Japan was jointly organized in Menomonee, Wisconsin, with the University of Wisconsin-Stout. The emphasis was on wood technology and related chemical utilization of waste as fuels and other by-products of the paper industry. Arising from this an in-depth investigation is being undertaken of the industrial potential of new technology in this area, sponsored by WINTECH, the technical arm of the Welsh Development Agency.

There is no doubt that higher education institutions and other public investments need to make better use of their resources to promote economic prosperity. In any event, in today's climate it is a policy of mutual self-help. A local example might well illustrate this. Clwyd Technics Ltd was set up to commercially exploit the educational microelectronics products of NEWI. Ed-Word is a product developed by teachers for teachers, to teach and exploit word processing in business studies. Producing the prototype product was easy compared with mounting a commercially effective production system for manufacturing the Ed-Word by the thousand.

Now a factory in Rhydymwyn (near Mold), employs 33 people and generates a profitable £1.35 million turnover to sell its indigenous products such as Ed-Word, Keymaster, Taster Kits, Tracker, and so on. Perhaps our own experience here strengthens our belief that we should ensure that other embryonic companies with sound innovative products do not fail because of a lack of marketing, management and production skills.

Science parks, research parks, and innovation centres

We conclude this chapter by profiling the individual developments taking place in the United Kingdom.

The location, date of establishment, telephone contact, management structure, and number of tenants are given for all existing parks, as well as a summary of proposed developments (Table 6.17).

Finally, it may be said that science parks, research parks, and innovation centres are different forms of the same phenomenon: combining a growing interest by academics in new technology-based firms, entrepreneurship, and small and medium-sized enterprises set in a context of profound structural economic change.

Table 6.17 Science parks in the United Kingdom, September 1986

University/ college	Name and address of development	Contact tel. no.	Partners	Date of opening	Total area (acres)	Buildings completed (sq. ft.)	Buildings under construction (sq. ft.)	No. of companies in park
Fully operational science parks (26)								
Aberystwyth*	Aberystwyth Science Park Cefn Llan Aberystwyth SY23 3AH	Conrad Jenkins 0970 615779	University College of Wales Aberystwyth Mid Wales Development	Feb. 85	6	12,000	10,000	5
Aston*	Birmingham Technology Ltd. Aston Science Park Love Lane Aston Triangle Birmingham B7 4BJ	Harry Nicholls Barbara Richards 021 359 0981	City of Birmingham Aston University Lloyds Bank Plc	1983	22	121,500	27,000	35
Birmingham*	University of Birmingham Research Park Vincent Drive Edge Baston Birmingham	Dr. Derek Burr 021 471 5977	Birmingham University Birmingham City Council	Apr. 86	12	31,000	—	8
Bolton*	Technology Exchange Spa Road Bolton Lancashire	David Bromley	English Estates Bolton Institute of Higher Education Bolton Borough Council	Oct. 86	2.5	25,000	—	10
Bradford*	Listerhills at Bradford University Bradford West Yorkshire	Lawrence West 0724 733466 Ext. 8284 Phil Wilbourn 0302 66865	English Estates Bradford University Bradford City Council	Mar. 83	11.75	64,700	10,074	24
Brunel*	Brunel University Science Park Uxbridge Middlesex UB8 3PH	Peter Russell 0895 72192	Brunel University	Apr. 86	6.4	46,000	—	9

Table 6.17 — continued

Cambridge*	Cambridge Science Park Milton Road Cambridge	John Tweddle Henry Bennett Letting Agents Bidewells 0233 841841	Trinity College, Cambridge	1972	130	528,000	50,000	68
Durham*	Mountjoy Research Centre Durham University Durham	David Rhodes 0642 765911 Dr Eric Howells 0385 64971	English Estates Durham University Durham City Council Durham County Council	Oct. 85	2.7	39,800	—	5
East Anglia*	University of East Anglia Science Park University Village Wilberforce Road Norwich NR4 7TJ	Philip Lannigan 0603 56161	University of East Anglia	Feb. 84	12	10,000	—	4
Glasgow/ Strathclyde*	West of Scotland Science Park Kelvin Campus Glasgow G2G 0SP	Alasdair McNicoll 041 946 7161	Scottish Development Agency Glasgow University Strathclyde University	Sept. 83	61.5	43,200	34,000	5
Heriot-Watt*	Heriot-Watt University Research Park Riccarton Edinburgh EH14 4AS	I. G. Dalton Director 031 449 5111	Heriot-Watt University	1972	56	240,000	30,000	18
Hull*	Newlands Centre at Hull University Inglemire Lane Cottingham Road Hull	Philip Wilbourn 0302 66865 David Geekie 0482 46311	English Estates Hull University Hull City Council	Dec. 84	3.0	25,000	12,500	12
Kent	Kent Research and Development Centre The University Canterbury Kent CT2 7NR	Bernard Watts 0227 66822	Kent University Kent County Council Canterbury City Council	Feb. 86	10	12,000	12,000	2

Table 6.17 Science parks in the United Kingdom, September 1986 — *continued*

University/ college	*Name and address of development*	*Contact tel. no.*	*Partners*	*Date of opening*	*Total area (acres)*	*Buildings completed (sq. ft.)*	*Buildings under construction (sq. ft.)*	*No. of companies in park*
Liverpool*	Merseyside Innovation Centre Ltd. 131 Mount Pleasant Liverpool L3 5TF	Arthur Rimmer 051 7080123	University of Liverpool Liverpool Polytechnic	1982	2	15,000	—	12
Loughborough*	Loughborough Technology Centre Epinal Way Loughborough Leicestershire	Roger Say 0533 87131 Loughborough Consultants 0609 230426	Leicestershire County Council Loughborough University	Apr. 83	3	22,000	—	16
London/ South Bank*	South Bank Technopark 90 London Road London SE1	Jeff Jeffers 01 928 2900	Prudential Assurance S.B. Polytechnic South Bank Technopark Co.	Apr. 85	2.4	50,000	25,000	20
Leeds	Springfield House at Leeds University Leeds West Yorkshire	Philip Wilbourn 0302 66865	English Estates Leeds University	June 83	1.96	32,200	—	11
Manchester*	Manchester Science Park Ltd. Enterprise House Lloyd Street North Manchester M15 4EN	Dr. Tom Broadbent 061 226 1000	Manchester City Council Manchester University Ciba-Geigy Plc Ferranti Plc Fothergill and Harvey Plc Granada Television Ltd.	Dec. 84	15.5	24,000	—	13
Newtech*	Newtech Square Deeside Industrial Park Clwyd CG5 2NU	Dr John Allen 0244 822881	Clwyd County Council The North East Wales Institute Welsh Development Agency	Dec. 85	Part of Dee-side Ind. Park	66,000	—	4

Table 6.17 — continued

Nottingham*	Highfields Science Park University Boulevard Nottingham NG7 2QP	John Webb 0602 506101 Ext. 3349 Tony Edwards 0602 418561 Ext. 354	Nottingham University Nottingham City Council	Dec. 84	14.0	31,000	24,000	14
St. Andrews	St. Andrews Centre North Haugh St. Andrews Fife	Marion Sherwood 0592 205171 Dr Morris Shepherd 0334 76161	Scottish Development Agency St. Andrews University	Dec. 84	0.74	12,000	—	1
Southampton*	Chilworth Research Centre Chilworth Manor Southampton SO9 1XB	John Stuart-Buttle 0703 767420 Prof. John Large 0703 559122 Ext. 2296	Southampton University Southampton Economic Development Corporation Ltd.	1984	26	46,000	30,000	15
Surrey*	Surrey Research Park University of Surrey PO Box 112 Guildford Surrey GU2 5XL	Dr Malcolm Parry 0483 579693	University of Surrey	1984	70	190,000	140,000	9
Stirling	Stirling University Innovation Park Stirling University Stirling	Ross Clark 0786 70080	Stirling University Scottish Development Agency Central Regional Council Scottish Metropolitan Plc	June 86	14	13,000	—	—
Swansea*	Swansea Innovation Centre University College of Swansea Singleton Park Swansea SA2 8PP	Sidney Brailsford 0792 295556	University College, Swansea Welsh Development Agency	July 86	3.4	30,000	—	4
Warwick*	University of Warwick Science Park Ltd. Barclays Venture Centre Coventry CV4 7EZ	David Rowe 0203 418535	University of Warwick Coventry City Council West Midlands Enterprise Board Warwickshire County Council	Feb. 84	24	102,000	52,000	30
Total (26 parks)					512.85	1,831,400	456,574	364

Table 6.17 Science parks in the United Kingdom, September 1986 — *continued*

University/ college	*Name and address of development*	*Contact tel. no.*	*Partners*	*Date of opening*	*Total area (acres)*	*Buildings completed (sq. ft.)*	*Buildings under construction (sq. ft.)*	*No. of companies in park*
Science parks in the course of construction								
Belfast Queens University*	Antrim Technology Park c/o IDB Regional Office Thomas Street Ballymena Co. Antrim	George Dillon 0266 3655	IDB University of Ulster Queens University Belfast	Oct. 86	80	—	14,000	1
Cambridge*	St. John's Innovation Centre Cambridge	Chris Johnston 0223 338627	St. John's College Cambridge	June 87	19	—	37,000	—
Cardiff	Cardiff Technical Centre Senghennydd Road Cardiff	Jeff Andrews 0222 499022	South Glamorgan County Council University College Cardiff	Oct. 87	1.5	—	Approx. 18,000	—
Keele	Keele University Science Park Ltd. Keele University Keele Staffs ST5 5BG	Dr. David Cohen 0782 621111	Keele University Newcastle under Lyme Borough Council Staffordshire County Council	Dec. 86	15	—	23,000	4
Total (4 new parks)					115.5		92,000	5

In recent years this phenomenon has grown at an exponential rate. It has moved from the esoteric to the fashionable. In the 1970s there was only the Cambridge Science Park; now there are 26, with four new parks in construction. These developments have been praised as key growth areas in a new economy based on 'sunrise industries' — a panacea for unemployment — but are often criticized as yet another tool of desperate regional policy. The truth probably lies somewhere between these two assertions.

In addition, a number of universities and other institutions of higher education and research are known to be at the initial planning/feasibility study stage of developing schemes, including: the University of Aberdeen, the University of Essex, Portsmouth Polytechnic*, Imperial College, Oxford University, the Medical College of St. Bartholomew's Hospital*, Sheffield Polytechnic*, Sunderland Polytechnic, Bangor University, Salford University*, and Reading University*.

*Members of the UK Science Park Association.

Notes

1 In the United Kingdom there are on average 29 service-sector (defined as construction, public utilities, transport, and communication) jobs per 100 population. Similar ratios prevail for other regions; for example in East Anglia there are 28 service jobs per 100 population. If any marginal increase in population is associated with the same number of service jobs as the average, a population expansion of 1,000 will result sooner or later in 280 local jobs. Clearly the precise effect will depend *inter alia* on factors such as the degree of excess capacity in public service provision and on the policy response of the local authority. In the case of the CSP the differential increase in population (compared with a development in the sub-region) is broadly estimated at 500 (for the 1,000 jobs on the CSP) and is associated therefore with perhaps as many as 140 jobs.

Part IV

European Community initiatives

Chapter seven

European Community initiatives

The ways in which technical and scientific research filters through to the industrial sector has in recent years become a hot political and economic topic in Europe. A host of people from politicians to university academics and industrialists are examining the routes by which this transfer takes place and techniques to make the translation more effective.

In Europe, technology transfer policies are seen as fundamental to improving economic growth. High-technology-based industries are the ones that will survive and prosper, creating wealth and employment and to some degree compensating for the decline in traditional European manufacturing industries such as steel, shipbuilding, and heavy engineering.

Individual European governments and the EEC have become particularly keen on technology transfer. They have taken a lead in pushing for definite economic returns, in the form of new processes and products, on the large sums of taxpayers' money spent on R&D.

Concern has grown that advanced scientific work carried out in Europe has often failed to translate into commercial success; optical fibres, video technology, and kits based on immunoassays for medical diagnosis are just three examples. Instead, in recent years the United States and Japan have raced ahead in many high-technology areas, frequently using the scientific breakthrough that the Europeans have pioneered. This explains the many initiatives taking place in Western Europe to narrow what it sees as the technology 'lag and lack' between itself and the United States and Japan (p. 190).

Mechanisms aimed at making better commercial use of scientific ideas include the Esprit and Eureka technology projects, in which individual countries in Europe are collaborating and pooling research efforts in disciplines such as electronics, new materials, chemicals, and transport (p. 192). Other efforts are focusing on universities as repositories of technical skills that need to be spread into the social fabric of a country to yield maximum economic benefit — hence the interest in science parks and technopoles (p. 195).

The European Commission has also developed several initiatives to cement stronger ties between academic institutions and the business community, particularly if they are pan-European (pp. 198–206).

Part of the development in European technology transfer in recent years has concerned efforts to steer the results of scientific research, and the uplift in economic fortunes it is assumed will accompany commercialization of the research, to regions badly hit by the decline in traditional manufacturing. This explains why a number of places hard hit by this decline (Dortmund in Germany, Deeside in Britain, for example) have set up their own technology transfer agencies to help existing companies become aware of new ideas that can improve products and processes.

Technology lag and lack

Throughout the last 30 years the European Community has proved phenomenally resilient and dynamic, and it has been so magnetic that it has seen its membership double from six to twelve states. The Community has grown from 169 million citizens to over 321 million citizens. Today it is the world's largest exporter with over 20 per cent of total world exports.

Its progress has been the result of political will and vision, supported by the efforts of Europeans to continue growth and development. Progressive dynamic policies and programmes are needed to ensure the future of Europe.

Few policy issues have preoccupied European politicians, industrialists, or academic researchers more intensely during the 1980s than the challenge of adapting to the new technologies and harnessing them to generate further economic growth and competitiveness. Anxieties about technology 'lag and lack' have surfaced with a vengeance in the past few years. Talk of a 'crisis' in Europe's technological performance has prompted a variety of actions by national governments and the EEC to try to stimulate a revival of competitiveness.

How justified are Europe's fears? Europe's expenditure on R&D, one common gauge of technological performance as we have seen in Chapter 1, is on a par with that of the United States and Japan. Of the $200 billion or so that the developed world spends yearly on R&D, the United States, Western Europe, and Japan divide up the cake in the rough ratio of 5:3:2, or put another way,

$560 billion is spent in Western Europe. Statistics show that each year Western Europe invests in R&D about 12 per cent less per head than Japan, and slightly more than half the comparable sum in the United States. Of greater importance than this is the fact that while Europe has been in recent years increasing its R&D spending at the rate of 4 per cent annually, expenditure in Japan has been increasingly at 10 per cent a year and at 6 per cent a year in the United States.

However, these figures hide who in Europe are the greyhounds and who the tortoises. Between 1967 and 1982, according to an analysis by Patel and Paritt of the Science Policy Research Unit at Sussex University, Ireland, Sweden, and Belgium all increased their R&D expenditure at a rate of more than 6 per cent per annum. France, West Germany, and Italy had reasonable increases over this period of 5 to 6 per cent a year, whereas the worst performers were Holland, Switzerland, and Britain — which managed annual R&D rises of just 1.5, 1.2, and 0.9 per cent respectively.

Furthermore, other statistics suggest that overall Europe gets less back in commercial returns from such investments than does Japan or the United States. For example, between 1970 and 1984 its total share of OECD area exports of high technology fell from 33 per cent to 26 per cent, while imports share of high-technology demand increased from 25 per cent in 1970 to 42 per cent in 1980. All such measures are, of course, subject to definitional arguments about what constitutes high technology, but clearly there is cause for concern.

But the decline in Europe's technological competitiveness is not manifested across the board. It has managed to hold its own internationally in a number of major industries, including pharmaceuticals, chemicals, and mechanical engineering. In aerospace and defence technology, it may still run a distant second to the United States but it is a good way ahead of Japan. The European Airbus and the Arianne space launcher are examples of Europe's success; and Europe can count some promising contenders in growth industries such as robotics, genetics, and new materials.

Where Europe undoubtedly has problems, however, is in electronics and information technology. In these sectors it has never attained a position commensurate with its economic importance, and its performance has continued to deteriorate. European suppliers account for a mere 10 per cent of the world microchip market and are minor league players in computing. The only electronics products in which Europe's external trade is roughly balanced are telecommunications equipment.

The heightened sense of concern in Europe about its technological future is attributable to three principal factors:

1. The sheer breadth and scale of the impact of information technology. Unlike any other technological advance this century, it is profoundly influencing developments in almost every area of industrialized society, shifting the pattern of business competitiveness and accelerating social change.

2. The growing perception of advanced technology in strategic terms. Uncertainties created by US export controls have persuaded several European governments of the need to maintain self-sufficiency in sensitive 'core' technologies, while President Reagan's Star Wars programme was viewed, initially at least, as likely to drain scientific talent away from Europe.

3. Increasingly fierce international competition has underlined severe structural handicaps to Europe's position. Protected home markets, differing standards, and the legacy of 'national champion' policies have produced a highly fragmented industry in which even the largest companies lack the economies of scale needed to compete effectively.

Europe seems to have lost the ability to innovate effectively and competitively.

All three factors have inspired a series of R&D initiatives from both the European Community and individual European governments.

Collaborative projects

In recent years the pace of technical change has accelerated and will continue to do so. In a number of areas such as micro-electronics, biotechnology, and telecommunications, even the biggest European companies have found it difficult to keep up with these advances using their own laboratory resources. Hence the moves by such companies as ICL in Britain, Siemens in Germany, and Electrolux in Sweden to strengthen links with university research groups, in a bid to bring to their existing operations new technical approaches. And the costs involved in R&D are now so high that few companies or even governments can afford to go it alone.

Esprit and Brite

Collaboration in R&D between European companies has emerged in the past few years as a major focal point of Europe's efforts to revitalize its competitive performance in advanced technologies. The most prolific source of initiatives has been the European Community, which has launched a succession of subsidized programmes, including Esprit in information technology and Brite in factory automation. Most participants have been enthusiastic about the achievement of these initiatives, intended to finance co-operation schemes for the pre-competitive research. Esprit has created the environment for further co-operation and the atmosphere of mutual trust, but it is not enough. It is one thing for the promotion of new technologies, but it is another to find the right market conditions.

Eureka

Eureka was created in 1986 and has been backed by 20 European governments. As well as embracing countries outside the EC, Eureka is intended to encourage the joint development of 'close to the market' projects with clear commercial applications. Whether this will happen in reality remains to be seen.

Race

The latest joint collaborative programme to be launched is Research in Advanced Communication for Europe (Race) with a definition phase started in December 1985 and 31 projects costing Ecu 40 million. The aim is to agree on a common basis for the telecommunication system in Europe for the future — wideband communication in which voice, data, visual, and text transmission will be integrated. The intention is to define the technologies to make broad-band communications a reality by 1995, bringing together state telecommunication bodies, manufacturers, broadcasters, and university researchers. Table 7.1 shows a sample of EC programmes involving collaboration.

Comett

Launched in 1986, Comett is a European programme of co-operation between universities and industry in training for technology. Its aim is to develop transnational training to help ensure the industrial and technological development of a unified Europe. The result is to have trained Europeans who together are capable of meeting the growing technological challenge and consequently will be able to meet the demands of world markets. Comett works by transnational exchange of students and personnel between universities and businesses; developing and initiating joint university–enterprise projects in the field of continuing education, particularly in new technologies; and supporting the development

Table 7.1 European Community programmes: a sample of major projects

Programme	*Date*	*Duration (years)*	*Budget (ECUs)*	*Emphasis*	*Future plans*
Bep	1982	4	15m	Promotion of academic/ industrial links; support of post-doctoral training and exchange.	Succeeded by Bap
Esprit	1983 (pilot phase)	1	11.5m	Microelectronics, software technology, advanced information processing, office systems, computer integrated manufacture.	Phase 2 (5 years)
	1984 (Phase 1)	5	750m		
Race	1985 (definition phase)	1.5	40m	Establishment of a technological base for the introduction of a Community-wide IBCN telecommunications infrastructure and service.	Phase 2 (5 years)
	1987 (Phase 1)	5	Currently being decided		
Brite	1985	4	125m	Development of advanced technologies to support traditional industries within the EC.	
Bap	1986	5	50	Promotion of research and training and contextual developments in enzyme, genetic and protein engineering.	
Comett	1986	4	80m	European network of university/ industry training partnerships.	Phase 2 (3 years)

Source: Claire Shearman in *European Technological Collaboration*, Chatham House Papers, February 1987, p. 36. London: Chatham House.

of multi-media training systems. In 1986 The North East Wales Institute of Higher Education (NEWI), in collaboration with local businesses and some from abroad, submitted an application to the European Commission for funding under the Comett programme. This programme is now proceeding successfully.

We have now formed a University–Enterprise Training Partnership (UETP) known as the NEWI-UETP. Its objective is to grow transnationally and to assist in the development of the region. A second objective is to become involved in other European initiatives recognizing the approach of the Open Market in 1992.

The NEWI-UETP has grown substantially and the partnership now includes three British institutes of higher education, 30 British companies, four European firms, and three European institutes of higher education. It is placing stress on developing further partnerships across Europe, and it is able to offer appropriate technological training regionally or across Europe.

Continental science parks and technopoles

Within continental Europe science parks are largely a child of the 1980s. Some, such as France's big complex at Valbonne-Sophia Antipolis, were in existence well before then, but there has been a big surge in activity in the last four years. For instance the first Nordic park, Oulu Technology Park in Finland, opened its doors in 1986.

Growth in the last few years has been rapid. France was early into the field and had three major technopoles, or technology parks, in the second half of the 1970s; it now claims to have at least 13.

A rash of openings has taken place in West Germany, including the famous West Berlin innovation centre Berliner Innovations und Grunderzentrum (BIG) and its next-door neighbour Technologie- und Industriepark (TIP). Over the past five years the Germans have opened about 20 parks, have planned more than a dozen others, and there are proposals for at least 35 more.

Nor does the list stop there. Belgium, also early into the field, has eight parks and plans for two others. The Netherlands has been relatively slow to develop parks and has only three. Italy is pressing on with a number of parks, hoping to bring high-technology companies and jobs to areas that are economically depressed, especially in the south. Greece is also well into the movement.

The thinking behind science parks is different in each country, and few have the same concept of technology transfer allied to university research that exists in Britain.

Many of the parks have grown very quickly and now claim large numbers of employees. At Sophia Antipolis there are said to be about 5,000 people employed, while Berlin's BIG has acquired over 30 companies (about half the number at Cambridge) and is now full. Some outside observers have doubts about how many of the claimed 5,000 people are actually working on technology transfer at Sophia Antipolis.

Across the rest of Europe there has been much more government input than in Britain, where apart from the early stimulation and encouragement of Mr Harold Wilson's Labour administration in the late 1960s, there has been relatively little official support. The private sector has played a relatively minor role in Germany, whereas the government has played an important one. The role of government in initiating science parks has been crucial. The development of technology-based firms is a central part of federal and regional government economic strategy and is very much influenced by government actions.

Financing of infrastructure for French technopoles has taken place almost entirely through the public sector. Although in some places there was initial interest in a technopole from local research on higher education establishments, the initiative has always been picked up by local authorities with the support of central government.

Elsewhere the story is similar. In the Netherlands recent growth of interest in science parks evolved partly from central government's increasing interest in regional planning. And in Belgium most of the science/research parks are developed through the local authority, usually the local regional development corporation.

The German interest in science parks has undoubtedly stemmed from a realization that the country was lagging seriously behind both the United States and Japan in the field of technological innovation. Japanese work in, for instance, watches (based to a large degree on British quartz technology) or US research on semiconductors, had left the Germans feeling that their strong engineering tradition and reputation were no longer sufficient in themselves.

The French had a more regional approach to new technology. They wanted to develop much of the southern coastline of France into a technological home for twenty-first century industries — a Silicon Valley from Toulouse to the Italian border.

Similar differences exist elsewhere. Belgium has consciously sought to attract research-oriented companies to its parks, and the

Dutch have sought local initiatives. While Belgium can now boast incoming companies such as Digital, Honeywell, Wang, IBM, and Monsanto from the United States, West Germany's Nixdorf and Japan's Terumo and Yamazaki, the Dutch have sought to develop the commercially exploitable activities of the universities.

Whereas the parks in Belgium support over 6,000 jobs, there are probably no more than 750 in Holland. The average size of a Belgian factory is about 70 employees (and much larger in the blue-chip multilaterals) that in Holland is around 10.

Belgium can claim with some justification to have been ahead of the game in recognizing the need for science parks. By the middle 1970s it had four: Evere-Brussels, Leuven-Aasrode, Liege, and Louvain. However, only the latter is really deeply involved with technology transfer and would really qualify for the appellation of science park. The park is next to the university, with close links to academic life.

Americans would be aghast at any atttempt to differentiate, as the British do, between science and technology parks, but in British terms the other Belgian parks are undoubtedly places where technological work of a high quality goes on.

The German parks look much more to technology transfer, though because of the big regional input they do tend to draw a much clearer distinction between those parks mainly involved in technology transfer and those where the prime emphasis is on the formation of new companies.

The German parks also have one other distinguishing feature: a large proportion of them have been set up in converted buildings. Only five of the established centres — Dortmund, Hamburg HTC, Heidelberg, Stuttgart-Pfaffenwald, and Syke — are in new, purpose-built premises. Even the newer, nearly completed centres follow a similar pattern. Of 14, only Bremen and Cologne are newly built.

Although Sophia-Antipolis is the best-known French park, with IBM and Hewlett-Packard among its tenants, it was not the first to be formed; that honour goes to Dirst de Meylan in Grenoble. This park now has about 150 tenants, which have created some 3,500 jobs.

The early French parks were all financed publicly and there has been a strong Paris commitment to the others, which now include Lyons, Bordeaux, Metz, Marseilles, and Paris itself.

The technopole is a much more diffuse concept than similar developments elsewhere in northern Europe. The early parks were relatively pure in concept and applied largely to areas with high scientific and technological content. It has become clear, though, that more recent developments have seen the technopoles act as a catalyst for regional development.

These differences with continental European parks have meant that the science park movement has developed in a number of different ways. To a large extent, the directions have evolved as a result of central government decision-making. Table 7.2 shows the dramatic growth in European science parks in recent years.

Table 7.2 Growth of the European science park movement, 1980–85

	1980		*1985*	
Country	*Parks*	*Establishments* on parks*	*Parks*	*Establishments* on parks*
Britain	3	76	13	180
West Germany	—	—	18	269
France	3	275	8	320
Belgium	4	38	5	76
Netherlands	—	—	3	42
Total	10	389	47	887

Source: Currie Sunman Partnership.

*Establishments include companies and research institutes.

European Business and Innovation Centre Network

In November 1984 representatives of a number of business and innovation centres from several member states of the European Community met in Brussels and decided to create an international association called the European Business and Innovation Centre Network (EBN) and to request Community support for this initiative. This support was granted by the Commission in December 1984, and the association has been legally registered in Brussels.

By November 1985 a group of 12 existing BICs had been formed into an embryonic network. Growth since then has been rapid and there are now 45 members throughout the Community. With the exception of Luxembourg every country has one, and GBN hopes to expand into Austria and Turkey. It plans to add another 60 to 80 centres over the next three years.

EBN is not a bureaucratic organization. While it expects to pump £100 million of the Commission's money into BICs over the next few years, it has an executive team of just six people in Brussels.

Business and Innovation Centres

The main function of a business and innovation centre is to stimulate entrepreneurship and improve the chances of success of new businesses with a view to additional and growing job opportunities. It initiates a process of permanent and deliberate creation of innovative activities in new or existing businesses of small or medium size.

Typically the companies they back will employ between five and ten people (they do not usually support lone ventures) and require venture capital of £50,000 to £100,000. The BICs that are usually linked to a nearby university provide technical, financial, and management advice as well as premises. A BIC should be capable of launching between 10 and 15 companies a year.

BICs are usually financed by local industry, the public sector and Brussels, which will meet half of the cost of centres set up in the EC's assisted areas. But the aim is to make the BICs financially self-supporting after five to seven years. They are expected to draw income from renting premises and royalties and minority shareholdings in their successful client companies.

Since the growth potential of the new businesses in most cases and in operational terms is determined by their technology content, centres can start by detecting and screening entrepreneurs (incubator centres of enterprise agencies) or by detecting and developing technologies. But in all cases a BIC has to be professionally managed as a business itself and provide comprehensive services and facilities. These include the search for, screening of, and development of both entrepreneurs and technologies, and training in basic entrepreneurial and management skills. This formula, which is leading to the emergence of a new profession, has proved to be effective in several countries during the recent years of structural change.

The economic significance of a BIC, over and above its direct impact, is to cause industrial and technological impulses and spillovers for the regeneration of the local and regional economy, thus acting as a germ-cell for industrial modernization, reconversion, and development. Creating a BIC can be the starting point for effective public–private partnership by pulling together the local potential of banks, universities, public authorities, existing enterprises, and human resources. They are an instrument of industrial policy that is more attuned to market forces than traditional subsidy-orientated policies.

European Network

The main function of the European BIC Network (EBN) is to promote, reinforce, and support business and innovation centres. To that end it will:

be a clearing house for the exchange of information and experience between members (BICs) as well as between their associated companies through an appropriate information system;

provide professional advice and support to the members (existing BICs) and potential members (BIC creation programmes) at reasonable cost through the promotion of best practices and an appropriate consulting structure;

encourage and facilitate co-operation between members (BICs) (and especially between the business connected to EBN members) on business opportunities, particularly in the field of marketing (e.g. joint ventures, agency agreements, etc.) and technology transfer;

provide services to the European Commission on a contracting basis.

Case study

Resdev, a small Halifax-based manufacturer of epoxy resin compounds used for factory floor covering, had always been put off the idea of exporting by the cost.

Unipro, a Dutch manufacturer of flooring compounds and adhesives based in Haaksbergen near Enschede, Holland, had been seeking a technology deal with a foreign company for some time.

In May 1987 the two companies signed a co-operation agreement. Resdev, which employs 12 and has an annual turnover of £1 million, will exchange its specialized knowledge of industrial floorings for Unipro's expertise in decorative finishes.

The British company gets a 3 per cent commission on any sales Unipro makes and is supplying Unipro turnover Fl 20 million [£6 million] and 32 employees with a compound that uses raw materials not available in the Netherlands.

Resdev was given a European push by the new Calderdale BIC, which is being set up in Halifax, while Unipro was prompted by the Bedrijfs Technologisch Centrum Twente, a four-year-old BIC based on the business and science park in Enschede.

Calderdale and Twente are part of a four-way data link-up that takes in two other BICs: Promotech in Nancy, France, and SOCRAN (Société de Création d'Activités Nouvelles) in Liège, Belgium. Ultimately all BICs throughout Europe will be linked.

Robert Frederic, head of SOCRAN, says the financial backing his organization has received from Brussels has been invaluable. 'EC finance allowed us to establish the proper links with Calderdale, Twente, and Nancy. We might have tried to do something like that on our own but it would have been a much more amateur effort.'

His counterpart at Twente, Gijsbrecht van Driem, echoes this view. 'EBN support helps us to get subsidies. It acts as a seal of quality and channels community funds to this sort of centre more quickly', he notes.

John Blackburn, head of strategy at the Borough of Calderdale, says the BIC framework, with its ultimate goal of financial self-sufficiency, gives a more commercial look to local authority initiatives. 'It's important we create the right forum to link with private industry and make local companies realize they are part of the European market.' (These comments were reported in the *Financial Times* (London), November 1987.)

Membership

No particular membership campaign has been undertaken. However, British founding members have contacted a number of British organizations since January 1985 with a view to their joining EBN. These steps have been successful. In the other EEC countries the other founding members have started to apply the same informal approach since the end of March 1985, as has the Secretariat.

The EBN has a two-tier membership. All members take part in all activities of the association, but only full members have voting rights. The founding members are: Promotech (Nancy), Heriot-Watt University Research Park (Edinburgh), British Steel Corporation (London), SOCRAN (Liège), NV Rede (Edmondhoven), Control Data Corporation (Minneapolis), Newtech Ltd (Clwyd), Business in the Community (London), BTC Twente BV (Enschede), SPI (Roma), Mr Baldini (Milano), IRI (Roma), CSATA (Bari), The Innovation Centre (Limerick). Presently a score of organizations have applied for membership.

Full members are Business and Innovation Centres

In order to confer a mark of quality on member centres and on the businesses created in or associated with member centres from the very beginning of the association, it was decided that the fulfilment of the following basic initial requirements should be pursued for membership of the network:

to aim at the permanent creation of at least 10 new businesses per year and thereby at the permanent creation of new jobs;

that the new businesses (or offshoots of existing firms) to be created should be based on a technology that gives them a lead over competitors in the market and include manufacturing and services to manufacturing (e.g. CAD/CAM), but exclude commerce; restaurants; trade crafts; and social, cultural, or sport associations or firms and most professions;

the intention of providing the full range of services and support required by start-ups and existing SMEs;

to provide appropriate vetting to keep down the failure rate (less than one in five);

to operate as a business itself.

Initially these requirements will not be imposed on the centres as mandatory in order to become a member of the EBN. They must, however, be willing to fit the requirements progressively as well as be open to recommendations.

Progressively, as best practices are being established, more stringent requirements might be proposed as to the organization and staffing of the Centres.

Full members

A.F.P. Acciaierie e Ferriere Pugliesi (Giovinazzo, Bari).
B.I.G. Berliner Innovations-und Grunderzentrum (Berlin) (to be confirmed)
Biotechnology and Innovation Centre Wales (Swansea)
Birmingham Technology Ltd Aston Science Park (Birmingham)
B.T.C. Twente-Bedrijfstechnologisch Centrum Twente B.V. (Enschede)
Cork Business and Technology Centre (Cork City)
CSATA Centrol Studi e Applicazioni in Technologie Avanzate (Bari)
Essener Technologie und Entwicklungs-Zentrum GmbH (Essen)
FILSE Finanziaria Ligure per lo Sviluppo Economico SpA (Genova)
Heriot-Watt University Research Park (Edinburgh)
High-Tech-Center (Hamburg)

Innovation Centre (Limerick)
LIFE (Motherwell)
Newcastle Technology Centre Ltd (Newcastle upon Tyne)
Newtech The North East Wales Institute (Deeside)
Nuova Italsider (Taranto)
PROMOTECH Lorraine (Nancy)
REDE N.V. (Eindhoven)
SOCRAN Société de Création d'Activités Nouvelles S.A. (Liège)
SITZ Saarbrucker Innovations and Technologie Zentrum (Saarbrucken)
Synergie (Thionville, Metz)
Technologie Zentrum Dortmund GmbH (Dortmund)
Tyne and Wear Innovation Centre (Tyne and Wear) (to be confirmed)

European venture capital

Venture capital is held by European politicians to be potentially important in attempts to catch up with the United States and Japan in the commercial exploitation of high technology.

Having seen how US venture capital has played a part in the rise of such important companies as Apple Computer and Wang Laboratories, Europeans are naturally keen to repeat the process at home.

The increase in venture capital witnessed in Britain in recent years is now spreading gradually, if unevenly, across the rest of Europe. According to the European Venture Capital Association (EVCA) the availability of equity finance for small companies in the community expanded by 30 per cent per annum to ECU 6.6 billion (£4.75 billion) in 1986, with increases ranging from 10 per cent in Belgium to just short of 120 per cent in Italy.

The EVCA report identifies some 244 venture capital providers, as against the bare dozen or so that existed in Europe in 1980. All this has been made possible by the growth throughout Europe of small company stock markets, providing the essential mechanism by which unquoted investors can take out their profits. Starting with the British unlisted Securities Market in 1980, and followed by counterparts in France and the Netherlands, some kind of public exit route now exists in most EEC countries.

Table 7.3 shows the equity finance for small companies in the EC, clearly showing Britain as a major source of equity finance.

The EEC itself has been active in promoting small company investment. In November 1986 community finance ministers agreed

Table 7.3 Equity finance for small companies in Europe

Country	*Estimate end 1985 (ECU millions)*	*Increase on 1984–85 (%)*	*Percentage of total*
Belgium	535.1	103	8.12
Denmark	127.2	393	1.93
Spain/Portugal	404.0	63.8	6.2
France	887.6	30.3	13.47
Greece	19.8	75.4	.03
Ireland	169.3	35.1	2.57
Italy	90.9	119.8	1.38
Netherlands	1,074.2	32.1	16.3
FRG	562.1	57.1	8.53
Britain	2,715.1	42.6	41.2
Total EEC	6,585.3	38.0	100.0

Source: European Venture Capital Association (EVCA).

to an ECU 1.5 billion loan programme for SMEs. However, at the time of writing the Brussels authorities' efforts to promote more innovative forms of finance now look as if they may be frustrated by the battle over the EEC budget.

The EEC's interest in risk finance dates back to 1980, to the publication of a study identifying a serious European shortage of finance for small technology companies, especially when it came to seeking funds to back international projects. 'The whole focus was to get managers to think European,' says Dennys Watson, head of the Commissions section for industrial innovation. 'This is the only way we can get small high-technology firms on to an equal footing with their counterparts in the US and Japan.'

The key to that, argued the Commission, was to find a way of encouraging Europe's venture capitalists to work more frequently in international syndicates. Starting with an initial meeting of 40 venture groups in Brussels in 1982, the Commission then agreed to back the formation of EVCA. So far, EVCA's main role has been more as a talk-shop than a marketplace for cross-border syndicates. As the association admits, well under a tenth of Community venture-backed projects are of a cross-border nature.

Nevertheless, a sign emerged that ambitious European ventures can get off the ground with the formation of European Silicon Structures (ES2) a $65 million pan-Community custom microchip start-up with operations in Britain, France, and West Germany. ES2 raised funds in six European countries, and is among the most visionary high-technology investment attempted in the Community.

The impact made by ES2 was also a reminder of the venture capital industry's most often heard complaint — that good quality technology investments are pitifully few. The truth is that Europe's universities and research establishments are teeming with bright ideas; the problem is that venture capitalists are too averse to risk.

Further EEC initiatives

European Association for the Transfer of Technologies, Innovation, and Industrial Information

In order to promote more transnational interchange in general between innovation advisers, the EEC is supporting an association called TII. The main aims are: to stimulate innovation in industry, to promote transnational technological transfer, and to encourage transnational co-operation between European companies. Its initial activities involved organizing professional exchange between members, short- and medium-term transnational secondments for training purposes, seminars on various aspects of innovation consultancy in SMEs, permanent study groups, and action aimed at forming a common market for saleable technologies.

TII's programme is as follows:

1. Creating a network all over Europe, of members who are experts in the field of technology transfer. TII now has 180 members, among them scientists, laboratories, consultants, entrepreneurs, Chambers of Commerce, and so on, who all have the chance to establish personal contacts wherever they want.
2. Organizing seminars and colloquia.
3. Organizing exchanges and group visits with the special help of the Commission of the European Community.
4. Building up and promoting Eurotech Link, a computerized and specialized communications system on technologies, business opportunities, and all kinds of industrial information.
5. Creating more working groups on special technology transfer aspects, such as innovation, information, licensing, and TII communication systems.

There is no doubt that TII will become a pressure group, a lobby in favour of technology transfer and industrial information in Europe, thus backing not only the European technology union, but the much needed political union as a whole.

Standing Technological Conference of European Local Authorities

Another organization receiving support from the community is the Standing Technological Conference of European Local Authorities (STECLA). The foundation of STECLA arose from an awareness that innovation is often hampered not by technical factors but rather by market conditions, particularly where the market is excessively fragmented. Its aims are as follows:

> better identification of local authorities' requirements for new or improved products to fulfil their responsibilities; a better exchange of information between members and all other public or private organizations in order to improve the efficiency and productivity of the services in their charge, particularly through the application of technical innovation.

Comparative Index of Norms in Europe

What, for instance, does a firm do when it wants to export its products to other countries in the Community in the face of standards that differ from those of its own country? (Efforts are admittedly being made to harmonize European standards, and the 'summit' in Luxembourg 1985 has confirmed the political will of the member states to move as quickly as possible in this direction, but, in the meantime, the sad reality of differing norms and standards has to be faced.)

One of the projects of the plan, called ICONE, is a 'Comparative Index of Norms in Europe'. This will enable enterprises or their advisers to identify without too much ado the changes, if any, that have to be made to products to make them acceptable in European countries other than those in which they operate traditionally.

Euro-Tech Alert

Another project that receives EEC support is called 'EURO-TECH ALERT: a European technology awareness scheme'. The aim of this project is to set up a co-operative technology awareness network at the European level based on British experience with its Tech Alert scheme, in which a specialist team from the Department of Trade and Industry conducts a systematic search from the reports produced by public research establishments to extract information likely to be of interest to industry. This information is then précised, and the results made available to the publishers of trade journals with a view to selection and publication.

Conclusion

As we have seen, techniques to improve the flow of scientific ideas from the academic environment to the world of industry have come under immense study all over Western Europe.

Industrialists, politicians, and local authority planners have all found they have an interest in assisting the transfer of these ideas, to bring about wealth-creating new businesses and bring novel approaches into existing industry.

In a pattern that is now extremely visible across Western Europe, local government departments and academic establishments are becoming keener to assist the mechanism of technology transfer by the vehicle of business information centres, science parks and technopoles, and collaborative R&D projects, and last but not least venture capital.

Finally, what is the effect of these mechanisms in aiding technology transfer? In many cases it is too early to judge overall results, although there are many successes, as we have witnessed in this chapter. Naturally, some of the initiatives will be more successful than others, but it is essential that every mechanism for improving technology transfer is tried if Europe is to narrow the technology gap between itself and the United States and Japan.

Part V

Evaluation and conclusions

Chapter eight

Evaluation and conclusions

Evaluation

The data in Chapter 1 indicate that Japan's spending in R&D is increasing year by year and in absolute terms is second only to that of the United States; meanwhile, expenditure in Britain is decreasing in real terms. This might please Japanese bureaucrats at MITI and AIST, but further analysis should give cause for concern. Japan's R&D is heavily oriented towards short-term industrial application. In fact, approximately three-quarters of the country's R&D spending, which in 1985 totalled ¥7,890 billion (£34 billion), is contributed by industry, while the comparable fraction in Britain is one-third. Only 15 per cent of the Japanese total budget is directed at 'basic' or pure research; this compares with 17 per cent in Britain. The accent on industrial needs is made even sharper by the low sums spent in Japan on military research, which in Britain accounts for 60 per cent of total R&D expenditure. In other words, 30 per cent of civil R&D in Britain is basic research, compared with 15 per cent in Japan. The emphasis on applied R&D may have been acceptable in the past, but if Japan's contemporary ambition is to achieve major scientific breakthroughs, then the emphasis may need changing. Japan is certainly making efforts to increase spending on basic research, but whether pouring on additional resources is enough to achieve the results they require remains to be seen. On the other hand, Britain would appear to have a better balance between pure and applied civil research, although this appears to have arisen more out of historical practice rather than positive planning. Moreover, in absolute terms Britain still spends less than the Japanese on basic research, and yet continues to spend vast sums on socially unproductive defence R&D.

In Chapter 2 we examined the organizations and establishments implementing R&D, and it was noticeable that the infrastructure was similar in both countries: namely universities, government laboratories, and industry. There is no need to repeat the statistics here but there is one area that should be highlighted. For some time now in Britain there has been an underproduction of scientists, engineers, and technicians. A major reason for this is that science and technology are no longer regarded as attractive careers and are unable to compete with the more lucrative careers such as business, law, management consultancy, and the financial services.

In the 1950s and 1960s the most sought-after jobs for scientists and technologists were in universities, research council establishments, government research laboratories, or similar establishments like the UKAEA, the Met. Office, and so on, with relatively autonomous work routines, life-long careers, and well-paid positions. Scientists in the 1980s tend to be low paid, tenure has almost disappeared, and they work in laboratories in both the public and private sectors with continually contracting budgets. This is hardly conducive to attracting our brighter people to enter a career in science or engineering. The opposite is true in Japan; science and engineering are almost a national obsession. Japanese scientists and engineers are well-respected members of the community and have equal status with other professions, often having jobs for life with their employer. This is perhaps why one in five university graduates major in engineering and in 1982 70,000 of them entered a career in manufacturing industry, a massive figure by any standards.

In Chapter 3 we studied British and Japanese industrial–academic interaction policies at both the national and local government levels, and in Chapter 4 we assessed the degree of interaction in human and financial terms.

In Japan the systems are complex and the initiatives have not been set up systematically; it was not an easy task to detail them logically and simply. Yet their obvious benefits to technological development are so impressive that the effort to identify them was, we believe, worthwhile. Industrial–educational interaction in Japan can be described as being in a transitional era, with every indication that it will henceforth be accorded the highest priority. In terms of the number of policies, systems, and initiatives, and the volume of finance flowing between universities and industry in Japan, the trend for interaction has expanded at a consistently high rate over the last 20 years.

However, a further development of industrial–academic interaction could not be made without a drastic change and improvement in government regulatory control. The Japanese government has tried to reform its regulations in recent years. For example, Mombusho, traditionally most conservative and bureaucratic in its attitude, has loosened some of its regulations governing practices in universities by allowing them to obtain funding outside of Mombusho. This would have been unthinkable in the 1960s, when professors were positively discouraged from consorting with industry. This new initiative has been welcomed both by industry and universities, but clearly it is not enough. On the other hand, MITI and STA have introduced far more effective policies than Mombusho, and moved faster and more flexibly. STA, for example, has introduced and successfully operated the Science and Technology Promotion System (Chapter 3, p. 70). MITI has created the next-generation Technology Promotion System (Chapter 3, p. 75). Whereas Mombusho's systems are just frameworks or guidelines for interaction, MITI and STA are introducing systems in which scientific targets are set and can be achieved by collaborative teams called from industry, universities, and national research institutes. Thus MITI and STA systems are more target oriented, more pragmatic, and accordingly more fruitful than Mombusho's initiatives.

Further efforts by the Japanese government, academics, and industrialists alike are being made in order to enhance the interaction between the university and industry. For the immediate future the present patchwork pattern of promoting this interaction is likely to continue.

In Chapter 4 we assessed the level of financial support of Japanese R&D carried out in industrial laboratories. The very high percentage must be the envy of the British government, but in Japan it might even be causing some anxiety. Industries of the twenty-first century, in 'knowledge-based' activities like artificial intelligence and genetically engineered pharmaceuticals, will need to feed off pure scientific research, which is more conducive to systems operating in universities and national laboratories.

To provide for the future of such industries Japan needs to strengthen its efforts in pure research and shows every indication of doing so. The current emphasis is directed toward scientific breakthroughs in pure science, and this ambition is enshrined in concrete at Tsukuba Science City. We profile this unique scientific conurbation in Appendix 1. It is the ultimate in science parks, with two universities and more than 50 research institutions transplanted to the new location. On the site of the Science Exposition of 1985, industry is also locating its prestige research centres. With this intensity of individual and collective effort, Japan has every reason to feel confident about its future scientific productivity. What may be lacking in inspiration is being counteracted with perspiration.

In Britain, industry and universities are now beginning to build up a partnership, as is evident from the many initiatives described in Chapter 3. Several factors have brought about a change in perceptions in recent years. Most dominant perhaps have been the government cuts in university grants, which dates back to 1981. One effect was to make universities feel they should be doing more to earn a living by striking up contacts with the outside world. The result has been a stream of action by universities and other institutions of higher education in setting up science parks, research parks, and innovation centres. It is, perhaps, too early to judge whether these initiatives are the best way for universities and industry to interact. This new fashion grew mainly from the experience in the United States and the Cambridge initiative in Britain. The initial results are encouraging. In 1986 a study carried out by Peat Marwick and commissioned by the United Kingdom Science Park Association (UKSPA) made an in-depth study of science parks, and we report their findings on p. 226, in particular the extent to which tenant firms interact with their local university. There can be little doubt that these developments have directed academic scientists' attention to the practical implications of their investigations, if not yet to full-scale commercial exploitation.

In Chapter 4 (pp. 99–120) we attempted to assess the financial interaction between industry and universities, but were hampered by the dearth of statistical information available in Britain, compared with our research findings in Japan. Furthermore, the small amount of information that is available is capable of much misinterpretation. For example, there seems to be a continual debate in Britain as to how much universities earn from outside sources. The government maintains that universities could do much more to raise money privately. According to DES statistics (UGC, *University Statistics 1985/86, Vol. 3: Finance*) universities earned £59 million or 2.5 per cent of their total income from industrial sources, and government ministers contend that this is not enough. University vice chancellors, on the other hand, believe the government to be misinformed about what universities actually earn and what efforts they make to earn that money. The misunderstanding may be due to the fact that governments compile their statistics from 'Form 3 Returns' to the UGC, which were originally devised as a system of accounting for public money. Although universities are now also allowed to return information concerning sources of income other than public, many sources of private income have to be left out. Income raised from consultancy activities, university-owned companies, conference business, and some science park income, do not show up on Form 3. Nevertheless, Table 8.1, compiled by the universities themselves, clearly shows that some universities are more active than others in raising money from

Table 8.1 The universities' income, 1985–86

Institution	*Total recurrent income (£ thousands)*	*Block grant (%)*	*Government research income (%)*	*Non-government research income (%)*	*Endowments, gifts, etc. (%)*	*Overseas students' fees (%)*	*Other services (short courses, etc.) (%)*
Aston	25,274	63.8	7.0	5.0	0.4	3.4	4.5
Bath	23,108	60.6	9.0	7.0	0.5	5.2	3.7
Birmingham	70,854	57.4	10.0	6.0	0.8	4.0	10.1
Bradford	28,852	63.8	5.0	4.0	—	6.6	4.8
Bristol	52,938	59.6	13.0	6.0	2.0	2.4	4.9
Brunel	23,189	60.4	11.0	8.0	—	4.2	2.5
Cambridge	84,732	54.1	15.0	8.0	2.6	5.6	1.4
City	24,600	53.8	5.0	3.0	1.1	9.6	7.8
Durham	29,357	64.8	9.0	4.0	0.9	2.4	4.1
East Anglia	24,713	64.4	6.0	6.0	0.2	7.5	1.4
Essex	18,266	51.9	7.0	7.0	0.3	14.8	3.5
Exeter	24,358	67.5	4.0	3.0	0.5	2.8	5.4
Hull	25,057	62.7	7.0	2.0	2.6	4.9	1.0
Keele	15,358	62.8	6.0	4.0	1.3	7.9	0.7
Kent	21,099	54.1	9.0	3.0	—	12.1	1.1
Lancaster	24,045	60.8	9.0	2.0	0.9	4.8	5.6
Leeds	73,988	61.4	8.0	6.0	0.2	5.2	7.8
Leicester	31,675	61.8	10.0	7.0	0.3	2.6	6.1
Liverpool	63,092	65.3	7.0	5.0	0.5	3.4	4.1
London Business School	6,407	35.5	4.0	16.0	—	3.6	26.8
London University	480,707	52.7	11.0	13.0	1.4	5.7	6.3
Loughborough	35,246	55.8	12.0	7.0	—	8.0	5.8
Manchester Business School	4,912	25.9	0.15	7.0	0.5	3.2	46.5
Manchester University	85,494	60.0	10.0	5.0	0.5	4.1	4.4
UMIST	35,824	50.6	13.0	11.0	0.4	10.5	4.8

Table 8.1— continued

Institution	Total recurrent income (£ thousands)	Block grant (%)	Government research income (%)	Non-government research income (%)	Endowments, gifts, etc. (%)	Overseas students' fees (%)	Other services (short courses, etc.) (%)
Newcastle	58,781	61.8	8.0	7.0	1.3	5.0	4.1
Nottingham	50,606	59.3	10.0	8.0	1.6	3.7	3.2
Oxford	97,744	47.5	18.0	10.0	6.0	4.8	2.5
Reading	35,016	58.0	9.0	5.0	0.4	8.1	5.6
Salford	28,880	49.2	7.0	7.0	0.5	7.9	17.4
Sheffield	53,471	65.5	9.0	5.0	1.0	3.7	2.7
Southampton	51,943	53.1	15.0	9.0	2.1	3.6	5.9
Surrey	28,273	50.3	12.0	9.0	0.4	6.2	11.2
Sussex	26,954	56.0	13.0	6.0	0.2	8.0	1.2
Warwick	37,349	54.8	11.0	6.0	0.5	4.7	10.5
York	21,245	58.3	14.0	6.0	0.7	3.6	2.1
TOTAL WALES	130,925	58.5	10.0	7.0	0.7	6.5	4.4
Aberdeen	41,745	61.5	7.0	4.0	1.3	4.6	10.2
Dundee	26,666	65.2	10.0	8.0	0.7	3.2	2.9
Edinburgh	86,847	59.2	11.0	5.0	1.4	3.0	7.8
Glasgow	79,527	61.9	8.0	3.0	1.7	5.8	5.8
Heriot-Watt	24,745	51.4	13.0	7.0	0.2	8.9	6.7
St Andrews	20,861	65.8	6.0	4.0	0.7	7.4	2.3
Stirling	16,678	61.3	8.0	3.0	1.0	4.2	6.2
Strathclyde	47,734	56.8	11.0	6.0	0.1	9.8	3.1
Queen's University Belfast	48,802	72.2	5.0	4.0	0.1	1.1	5.7
University of Ulster	35,130	79.8	3.0	2.0	—	0.8	1.3
TOTAL UNITED KINGDOM	2,379,064	57.8	10.0	8.0	1.2	5.3	5.5

Source: University Grants Committee, University Statistics 1985–86. Volume 3: Finance.

private sources. Another estimate was given in the *Financial Times* (30 October 1986), which reported that in 1984–85 British universities earned directly from industry approximately £48 million, a 46 per cent increase on the £33 million earned in 1983–84. This trend has continued as the pressure applied by government has intensified.

Table 8.2 The universities' income, 1984–85

University	*Recurrent income, 1984–85 (£ millions)*[a]	*% from UK industry*	*University*	*Recurrent income, 1984–85 (£ millions)*[a]	*% from UK industry*
Salford	25.5	11.91	Wales	123.4	3.15
Heriot-Watt	21.7	8.69	Southampton	46.6	2.97
Loughborough	32.0	8.28	Leeds	68.2	2.87
Birmingham	66.7	7.04	East Anglia	23.1	2.80
Surrey	25.7	6.95	Bristol	49.7	2.57
City	22.6	6.14	Cambridge	76.8	2.51
Brunel	21.3	6.02	Lancaster	21.7	2.37
Essex	16.1	5.80	Oxford	88.0	2.33
Bradford	24.0	5.37	Manchester[b]	110.7	2.16
Strathclyde	43.6	5.37	Exeter	25.2	1.98
Reading	33.5	5.14	Glasgow	73.9	1.89
Aston	25.2	4.91	Sheffield	48.7	1.77
Stirling	14.9	4.39	Edinburgh	81.0	1.77
Leicester	29.3	4.00	Aberdeen	38.4	1.55
Durham	26.4	3.98	York	20.1	1.33
Warwick	30.4	3.73	Hull	21.7	1.02
Bath	21.1	3.59	Liverpool	58.6	1.02
Nottingham	45.5	3.58	Queen's,		
London	453.4	3.55	Belfast	43.5	0.66
Newcastle	53.3	3.43	Keele	13.9	0.65
Dundee	24.4	3.25	Kent	19.0	0.58
Sussex	25.6	3.22	St. Andrews	18.4	0.53

Source: Financial Times, 30 October 1986.

[a] Research for UK industry plus income from special and short courses.

[b] Includes University of Manchester Institute of Science and Technology.

Table 8.2 gives the *Financial Times* estimate of income earned from industry. It is difficult to reconcile the figures, and there is clearly a need to improve the statistics in this area.

In Chapters 5, 6, and 7 we examined several case studies and examples of industrial–academic interaction, selected according to generally regarded good practices.

It is impossible to draw conclusions from such a wide variety of case studies. Nevertheless, many academics, industrialists, and government officials in Britain, Japan, and Europe are now convinced that interaction between industry and universities is productive and that benefits will accrue to both sides by increasing the area of overlap.

The benefits accruing to universities from interacting with industry can be identified as follows: research and academic courses are more relevant; academics become more aware of industrial problems and practices; students' employment prospects are enhanced; staff development is improved; revenue is generated; ideas are generated (university research and student projects); equipment and materials are sometimes provided by industry; university credibility is enhanced by good links with industry; and the university has opportunities to apply theoretical knowledge to real-life practical situations.

The advantages to industry of interacting with universities are as follows: universities provide a pool of skilled labour/resources (e.g. libraries, computing facilities, particularly relevant to small- and medium-sized enterprises); universities can be a source of inexpensive research; universities offer a multi-disciplinary approach; universities can provide R&D facilities for small- and medium-sized firms; universities act as independent arbiters; universities can help industry to speed up/enhance the process of technical change; universities provide access to government funding; universities are a rich source for staff; universities provide access to data not normally made available by competing industry, including international data; interaction with university can enhance a company image; and there are synergistic benefits from creating a positive environment for interaction of industries and research personnel.

In Britain, although the provision of inexpensive research expertise and facilities appears to be an advantage to industry, it could be a disadvantage to the universities. The effort could also inhibit basic research by consuming the available creative manpower. Moreover, if research is carried out cheaply in universities, it may tend to devalue the research in the eyes of industy. New guidelines are being drawn up that will require industry to contribute more towards the real costs. As always, industry complains that this trend will hamper collaboration. However, with the move to establish more powerful research centres in universities, industry can be guaranteed a better service. Contracting out specialist needs in finance, advertising, transport, and so on is now well established in industry. Research will surely follow if universities can manage their services professionally and provide value for money.

In those areas where research can best be termed speculative, the university system offers a means of spreading the risk. Using the university in this fashion means that expensive company resources are not tied to projects with a low probability of yielding 'a marketable product'.

There are disadvantages in the interaction between the partners, but often these are a matter of perception rather than fact. One relates to a type of work undertaken that might reduce the freedom of academics and trivialize their research. This appears to emanate from those universities where the links with, and possibly the understanding of, industry are not great. The majority view by far is that university–industry collaboration should be encouraged.

The case studies show that the main barrier to developing university–industry interaction is one of management and structure of the link. Many of the disadvantages, problems, barriers, and disincentives stem from the rigidity and obsolete structures within the universities themselves. For example, the lack of staff time to undertake industrially oriented research is a key barrier. This can be overcome by managing existing resources more effectively and reorganizing present structures and possibly increasing research staff numbers. The present dual teaching-research system controlled by the all-powerful professor must surely now be questioned. Often small departments, based on a narrow discipline, overburden creative young people with teaching and administration. The professor alone determines the emphasis, creates the rules, and enforces them. The trend to build up larger multi-disciplinary units with a broadly based management can allow individuals to allocate more of their time to research, according to the resources they can attract from within and outside the university. Already minds have been concentrated by the financial necessity for 'rationalization'. Performance assessment too will enable the resources to be allocated to the more productive faculty members. The objective surely must be to loosen the bureaucratic bonds that now frustrate young scientists in their most productive years.

More attention must be given to the mechanisms of industrial contact. Industrial liaison units are generally understaffed. Specialist units should also have their own liaison staff, since project identification and planning is a highly expert operation. The problems (and/or barriers) of career structure, confidentiality, and areas to be tackled related also to the publication of the results of any joint research undertaken with industry. The need to publish is vital for such research to be recognized by existing university promotions committees. These should also seek to resolve how work with industry could be recognized for promotion purposes. Financial incentives too could provide a stimulus.

The following are some of the problems encountered by industry when interacting with universities: problems of making contact, and identifying the particular expertise; mismatch of industry/university objectives (e.g. meeting production deadlines); project management difficulties (many industrialists have experienced real difficulties in ensuring that a co-operative research project proceeds smoothly from conception to completion); negative attitudes by academics to certain types of applied research; confidentiality difficulties (this should not be a problem if certain safeguards such as patents are adhered to); ownership rights; contractual and institutional organization; geographic difficulties. Finally, an issue frequently raised is whether the new emphasis on forging industrial links with academic institutions could distract such bodies from what many consider should be their overriding mission of general teaching and the pursuit of knowledge.

There is also a fear that any efforts by the universities to obtain more money from industrial sponsors will give the government an excuse to reduce further the level of state financing. For many academic bodies, industrially oriented research schemes are still in their infancy. Universities as a whole, it appears, still have some way to go before case studies of the sort described in Chapters 5 and 6 begin to outweigh the more conventional operations of academia.

It is important that the industrial–academic liaison be developed steadily over a period of time. The present precipitate rush to seek industrial support in British universities has also its pathetic aspect. Staff, for years nurtured on the ivory tower doctrine, cannot overnight discard their upbringing. Those suitable for the new function should be carefully selected, suitably trained, and supported by appropriate structures within the university. Moreover, the emphasis on basic research must not be diluted, since this, not applied research, is the main expertise of universities. Japan has fully recognized this fundamental truth, and is moving to further increase the resources available for this function. Britain would do well to heed this lesson.

Japan: conclusions

In the early days of the research on semiconductors and transistors in Japan, very few scientists could understand why germanium crystals revealed amplification when two fine needles were set on the surface of the crystal. We learned about the science of semiconductors through scientific journals mostly from the United States, such as *Bell System Technical Journal*, *RCA Review*, *Journal of Applied Physics*, etc. We were extremely

> excited to do research on this unknown material, and worked very hard.
>
> The government laboratory we worked in was just a simple wooden building and the temperature inside was almost the same as that outside of the building. We had no central heating and, of course, there were no air conditioning systems. That was the condition of our laboratory in late 1940s.

This statement by Dr Makato Kikuchi, Managing Director, Sony Research Centre, illustrates how far Japan has progressed in such a short time. His full paper (Appendix 2) is worthy of detailed study, since it demonstrates that the perception of Japanese attitudes in the west does not accord at all with the reality of the situation. When Dr Kikuchi left the MITI laboratory to join Sony in 1974 he entered the 'battlefield' of the electronics industry. If not truly a war, it was certainly a race with the west in which Japanese industry needed to 'catch up', and the contest was waged under war-like conditions for the Japanese population. Personal ambition and gain were submerged in the mammoth national effort. It was vital to eliminate the 'Made in Japan', label, which was equated with low price and low quality during and immediately after World War II. Lack of indigenous resources had to be overcome by obsessive personal commitment and effort. The education system too was geared towards supporting this 'catching up' process. Japan took upon itself the role of the 'second climber', who could exploit the research already undertaken by the pioneers. Unjustly, perhaps, in this way they gained the reputation of copying rather than innovating.

Creativity

We have no difficulty in acknowledging the inspiration and creativity of the poet when he relates previously unrelated facts or situations in his ode or sonnet. The Welsh poet R. Williams Parry was a master in achieving this synthesis in his sonnet 'Tylluanod':

> Pan lithrai gloywddwr Glaslyn
> I'r gwyll fel cledd i'r wain

He relates such quite different experiences as the flowing of a mountain stream and the slick sliding of a sword into its scabbard. This interpretation and inspiration is a creative art. It was Max Planck who said in *A Scientific Autobiography*, William & Norgate, 1951, p. 109) that 'science was a created work of art' because 'When the pioneer in science sends forth the groping feelers of his thoughts he must have a vivid initiative imagination, for new ideas are not generated by deduction, but an artistically creative imagination'. It is disastrous that science has been described as if it were dull deduction from the facts.

Fleming worked in St. Mary's Hospital, Paddington, surrounded by the dust and grime of London. Spores of penicillin mould must have regularly contaminated bacterial colonies within the hospital. The technician must have regarded this as a nuisance that invalidated his experiments. Fleming, however, related two situations that had previously remained apart. The area around the contaminating mould was clear of bacteria. Here was, therefore, a potentially new material for selectively killing bacteria. Whether the discovery actually occurred exactly in that way is of little consequence. It is sufficiently close to illustrate what Bronowski has called a 'mental leap' leading eventually to penicillin and the other broad spectrum antibiotics. The scientific discovery requires an actively creative imagination. It is not surprising that Bronowski commented, 'The layman's key to science is its unity to the arts.' It was also this relationship that led Sir William Rowan to describe Laplace as the 'Shakespeare of mathematics'.

Creativity in science, if the view advanced is accepted, must, therefore, arise from the total development of the human personality, and be deeply rooted in the basic educational process that nurtures this faculty.

The development of the Japanese educational system after the war was directed more towards economic progress rather than development of the individual as an independent thinking and acting personality. The emphasis has been placed on the development of the group, and nonconformity was discouraged. Ever since the Meiji era the objective has been to 'catch up' with western civilization. Education was the way to achieve this. Obtaining a place in a good school would be the passport to a good university, leading to a good job. The competition was intense within the overcrowded islands. In 1984, 94 per cent of the middle school population graduated to high school and 30 per cent progressed to university.

This vast educational machine has produced a different type of scientist and technologist from the west. It had been fashionable to conclude that they are less 'creative' than their western counterparts. Dr Kikuchi in Appendix 2 and others I have talked to contend that it is a different kind of 'creativity'. The process described above, essentially a personal activity, is defined by him as 'independent creativity', which opens up an entirely new field (p. 247). The Japanese 'creativity' he describes as 'associated creativity', not really target-finding, but target-oriented. It starts with hints that are already available. Whereas the transistor was invented by the three scientists Bardeen, Shockley, and Brattain, the 'creative' introduction of the transistor into consumer products required 'associated creativity'. Its features are target orientation, action, co-operation, modification, improvement, and systematization. Western scientists might well regard this as development.

Whatever the definition, for progress to be achieved the basic discovery or 'hint' is necessary, and there is a general acceptance that the Japanese system in the past has not been as successful as the west in this regard.

Industry looks towards the universities to provide this inspired thinking, and the evidence previously advanced shows that ever increasing resources are being invested in basic scientific research in Japan. The 1988 White Paper on Japanese science points to the large volume of research published and the growing number of international patents filed as productivity and creativity in Japanese laboratories have increased. The country's substantial research budget has also now surpassed the ¥8 trillion mark (£35 billion), and is second in size only to that of the United States. The push for greater creativity is reflected in the wider variety of research projects being backed by the new science budget. Superconductivity, AIDS research, nuclear fission, biotechnology, and the development of new materials are all part of a growing commitment to creative basic research. The continuing internationalization of Japanese science is another major theme highlighted in the White Paper. Japanese participation in international exchange projects is increasing and more funds are being allocated to foreign scientists wishing to conduct research in Japan. The government hopes that the presence of foreign scientists will provide further impetus for new ideas.

Creativity is also being encouraged by changes in high school science lessons. Teachers are now asked to place more emphasis on the development of investigation and critical thinking skills rather than the rote learning and memory work traditionally demanded. A more stimulating atmosphere is detected in the National Research Institutes, universities, and company laboratories that carry out much of Japan's best R&D work. The onus is on the universities to catalyse this change. Professor Masamitsu Ohta, President of the Nagoya Institute of Technology, (Appendix 3) illustrates in his paper the importance of this activity within the overall historical context.

Evaluation and conclusions

Higher education–industry co-operation

All the data presented illustrate the increasing pace and complexity of the joint programmes and various interactions between higher education and industry. The government is supporting the overall objectives, but various traditionally bureaucratic departments are finding it difficult to relax their grip. University scientists, and indeed all staff, are in effect civil servants. The university administrators obtain their orders directly from Mombusho and not from the university president, as in Britain. The room and speed of manoeuvre of the head of the university are thus constrained, particularly since even within the university the Japanese consensus system prevails. Consultation is obviously good, but it is slow, cumbersome, and contrasts strongly with the benign dictatorship operating in the United States. The external agencies such as STA and MITI have, therefore, been more successful than Mombusho in achieving industrial collaboration. The experience of MITI and STA in managing scientific resources and programmes has allowed them to set up collaborative teams from industry, universities, and the national laboratories. Mombusho, on the other hand, has not been pro-active, and generally only drew up guidelines within which industry and universities might operate, if they so desired. Other ministries, such as Welfare, which covers mainly the health sciences, and the Ministry of Agriculture have been even less effective than Mombusho in initiating joint programmes. Even so, all have declared their intention to move positively.

As in other countries, the greatest handicap is the differing perception about the value and objectives of the research between universities and industry. Targets and end-product are the natural framework for industry, whereas the universities have a longer time scale, and are geared to achieve peer recognition through publication and participation in international conferences. Whereas in Europe the funding of research from outside the university is becoming a necessity, the average Japanese professor is able to maintain a core group of junior staff and graduate students to support his or her research objectives.

There is a difference in effectiveness too between large and small firms. The former possess the resources and knowledge to identify their likely partners in universities. The latter are mainly concerned with short-term solutions for the technical problems they encounter.

Overall, as noted, great progress is being made in building up co-operation. There are, however, complaints that continually surface:

> The research topics provided by industry, particularly small firms are not scientifically attractive and are too short-sighted (university professor).
>
> Research workers at university are mainly inexperienced postgraduate students. Industry requires post-doctoral expertise for their problems (a large firm).
>
> Great problems are encountered in identifying the appropriate contact to collaborate with (a small firm).
>
> Industry provides only a small amount of money, but expects enormous returns (university professor).

Despite the difficulties now encountered in building active industrial–academic partnerships, the trend is towards greater activity. The status of the professor will ensure his or her continued involvement with students after they leave for a career in industry. The annual New Year pilgrimage back to his or her alma mater by the industrial scientist is possibly the most effective contact point between the two sides. Thereafter, they are able to draw upon the increasing range of programmes for continuing their association. The distinguished professor, too, is continually sought after by companies to offer advice, arrange specialist symposia, and familiarize their scientists with new techniques and ideas. Industry is hungry for new ideas, and this will continue to fuel the industrial– academic machine.

Appendix 2 describes the response of one famous technological institution to the growing university–industry collaboration. Professor Ohta identifies the key to successful joint programmes as effective co-ordination and management of research. Mombusho alone has increased its sponsorship of joint industry–university programmes from 56 in 1983 to 250 in 1986 at a cost of ¥1,800 million. The national universities alone obtained ¥22,400 million from such joint projects. Industry too is increasingly undertaking major investment on university campus, such as the Advanced Telecommunications Research Institute at Nagoya.

After a rather halting start, therefore, the present position is one of dynamic, albeit uncoordinated growth in the collaboration between industry, universities, and the national laboratories. On both sides of the boundary the climate is conducive towards co-operation. As the value accrues to one side or another, it must be expected that preferred routes will emerge. It is not the normal Japanese style of planning, but its very diversity may in this instance prove advantageous.

Britain: conclusions

The growth and nature of British Science Parks

The public sector has to date invested £57 million in property developments for science parks. The majority of existing schemes are intended to promote the formation and growth of small 'high technology' firms. With some notable exceptions, the ethos is the promotion of small entrepreneurial private companies; it is an economic or business rationale for the use of public investment, rather than any social or collective purpose.

In brief, science parks are up-market industrial estates intended to house 'high technology' companies. What separates them from industrial estates is that some involvement by a university or other higher educational institution (HEI) is fundamental. In practice this involvement varies from park to park, but the underlying principle behind science parks is that HEIs represent a resource, in terms of knowledge and research, that may be tapped in order to promote 'high technology'-based growth in the economy.

There are many types of 'up-market industrial estates', which have a bewildering array of titles from 'technopark' to 'business park' to 'enterprise park'. One 1983 report identified 60 science park–type schemes (Debenham, Tewson, and Chinnocks 1983), and Henneberry counted 143 'high-tech property schemes' existing or proposed up to October 1983 (Henneberry 1984 in Peat Marwick Survey). Only a minority of these would have any claim to an HEI link. How science parks are defined thus determines how many developments may be counted as coming under that general term.

In view of these definitional problems it is convenient to focus on the parks that have joined, or are eligible to join, the UK Science Park Association (UKSPA). This organization includes the majority of HEI-based parks, including Newtech (Clwyd) Ltd, and has a published set of criteria for eligibility for membership:

> The Science Park is a property-based initiative which:
>
> has formal operational links with a university or other higher educational or research institution;
>
> is designed to encourage the formation and growth of knowledge-based business and other organizations normally resident on site;
>
> has a management function which is actively engaged in the transfer of technology and business skills to the organization on site.

According to UKSPA there are currently 28 operational science parks (see Chapter 6), with seven more under construction. A further five HEIs and one research institute are at the planning/feasibility study stage. These figures include parks that are not UKSPA members but conform to the above criteria (four of the 28 operational parks are non-members). These data exclude parks such as Wavertree Technology Park in Liverpool, where the Plessey company has been involved in the role of host institution, and Birchwood Science Park at Warrington, which has some links with neighbouring British Nuclear Fuels Ltd.

There are 412 tenant companies in the 28 operational parks. Of these 68 firms are in the Cambridge Science Park (the largest) and 144 are on parks in the 'southern sunbelt' (including Cambridge). UKSPA claims a tenant failure rate of 2.8 per cent per annum (12 firms), which is around one-third the rate for the whole small-firms sector. UKSPA puts the total employment figure for all 28 science parks at 5,300.

The total investment in land and premises to date in the 28 parks is £92 million, with £47 million planned for expansion, and £14 million for new parks under construction, according to UKSPA. The public sector is by far the major investor, contributing 62 per cent of the total.

There is a strong regional variation in public-private investment. Leaving aside the contribution of the host HEIs, which for the most part does not involve financial investment, in the 'southern sunbelt' only two parks out of six, Southampton and Kent, have public-sector investment; outside the sunbelt only three parks out of 22 do not have direct public-sector investment. Of these three, the Merseyside Innovation Centre received some help from the former county council, and Heriot-Watt has had new buildings financed by the Scottish Development Agency.

There is currently around 1.9 million square feet of property on the 28 existing parks. Of this 28 per cent (530,000 sq. ft.) is on Cambridge Science Park. Of the total available premises 11 per cent are empty at present, with some locations experiencing greater difficulty attracting tenants than others; Aberystwyth has been particularly slow.

All parks are different, making broad generalizations difficult. In addition, apart from CSP and Heriot-Watt, all British parks are less than four years old, and thus evaluation could be premature.

However, a 1985 study (Currie 1985) provided data which showed that less than half of firms in science parks were new; 53 per cent were either relocated firms or subsidiaries of multi-site corporations. This varies considerably between geographic locations, however. For England except the south east, the percentage of new firms was 54 per cent. On individual parks, Cambridge had a low level of new firms at 28 per cent, whereas Aston had a high level of new firms at 69 per cent. Warwick, Bradford, and South Bank had 50 per cent new firms. In addition to Aston's strong policy of promoting start-up firms, the park also has its own venture capital fund (funded by the city council and a clearing bank; this facility fills a gap for risk capital in Birmingham).

The formation of new firms in a science park does not, of course, demonstrate that these firms would not have been established anyway, without the presence of the park.

The model of science park growth suggests that university academics will commercialize their research products by establishing firms on the park. In practice, numbers of spin-off firms formed by the universities of HEI academics appears to be fairly low. Currie (1985) shows that 24 per cent of 187 firms surveyed in British parks were HEI spin-offs. Cambridge Science Park firms included only 11 per cent university spin-offs, but Aston has 37 per cent university spin-offs.

The 1986 UKSPA/Peat Marwick McLintock survey

In 1986 the largest ever survey of firms currently located in science parks took place. It was initiated by UKSPA and co-ordinated by KPMG Peat Marwick McLintock and it addressed the question of estimating the 'added value' of science parks. Firms were asked to provide indices of their performance, notably their growth in employment, output, and profitability. They were also asked for their views on the benefits they obtained from a science park location and their relationship with the host university/HEI. Below is a summary of the important findings and conclusions reached from the survey.

Linkages with universities/HEIs

The 1986 UKSPA survey revealed that 60 per cent of science park firms have used informal HEI contacts. Twenty-eight per cent of firms used consultancy or part-time HEI staff, and 14 per cent sponsor HEI research.

Further investigation (Wield and Quintas 1986) suggests that research links with the host HEI are usually only found in firms that have pre-existing links with the HEI. In many cases even these decrease over time. Firms that move on to science parks with no previous HEI research links do not usually make those links. Clearly it is easier to 'tap out' than to 'tap in' to the HEI.

The survey also showed that more than twice as many science park firms thought the general 'prestige and image of the site' to be an important factor in their choice of location than those mentioning the prestige of the university. Less than half the science park firms thought 'access to HEI facilities' to be an important consideration in their choice of location. Also, northern firms in a science park have stronger formal contacts with the university than southern park firms. Additionally, northern park firms place greater emphasis on the use of HEI facilities.

Employment trends and success in letting property

Current employment in British science parks is around 5,300, an increase of 53 per cent over the 3,800 recorded in 1985. This gives an average of 12.9 employees per firm, or 10.4 per firm if the five exceptionally large firms on CSP that employ over 1,000 people are deducted. The 5,300 jobs result from an investment in property alone of £92 million — £17,360 per job.

The survey also demonstrated that the small independent firm in the 'high tech' sectors exhibits significantly faster employment growth than a firm of comparable age in the more conventional sectors. Employment creation of high-tech firms located in the north is, however, significantly inferior to those located in the south. In addition, evidence suggests that businesses owned by academics and former academics appear to exhibit slower employment growth than businesses owned by 'professional' businessmen.

UKSPA survey data from 1986 show that there is a high proportion of qualified scientists and engineers (QSEs) among science park personnel; 44 per cent of employees are QSEs. Science park advocates would argue that it is still early — jobs will follow as firms grow.

Although 11 per cent of science park premises are currently unlet, the success of parks in some areas to let units at above-market rentals, in locations where there are acres of unoccupied commercial premises, is surprising. Aston is a notable example. The West Midlands had 30 million sq. ft. of empty industrial and commercial property in 1984, but Aston Science Park has been able to fill its new and refurbished premises at rentals above local industrial rates.

Sectoral concentration and technological level of firms

The 1986 UKSPA survey reveals a heavy concentration of firm activity in the computing, microelectronics, and electronics sectors; 53 per cent of firms are in these sectors, not including consultancy firms. Of all firms 34 per cent are in computing-related activities, and 23 per cent are in software. The remainder of firms are spread evenly over pharmaceutical, medical, mechanical engineering, environmental, and technical/business services.

Concentration in electronics and computing sectors is attributable to the low cost of entry into these sectors for firms applying and using existing technology. Whereas setting up to manufacture microchips is extremely expensive (around £100 million for the next generation), the use of chips in applications is low cost. This has several implications: reliance on largely foreign-owned manufacture; over-concentration within one sector, where the bandwagon may have rolled out of sight, to the possible detriment of alternatives where 'leapfrogging' may be possible; the implications of microelectronics applications for the wider economy, the possibility of 'jobless growth', and so on.

In the 1986 UKSPA survey, 31 per cent of science park firms considered their product to be 'leading edge technology', and a further 42 per cent stated that their products or processes involved the application of advanced technology. Interestingly, there was no significant regional variation in firms' perceptions of their technological level.

Of the firms 91 per cent said that they were involved in R&D, and one-half of the sample claimed to spend over 10 per cent of turnover on R&D. Relatively high percentage R&D expenditure may, however, be due to the newness of firms, often with low turnovers. A very high percentage (34 per cent) claimed that their R&D was in the area of 'radical new research' (Wield and Quintas 1986). These interview responses must be treated with caution. In-depth firm-level studies suggest that the majority of science park firms are involved not in radical innovation, or even any kind of innovation, but are in the business of applying existing technology.

This presents something of a paradox. On the one hand, lack of innovation would represent a failure to achieve the promise of much of the rhetoric about science parks and the interface with university 'leading edge' research. On the other hand, innovation itself does not have significant immediate economic effect; it is the diffusion of new technologies that has large-scale economic consequences. Thus science park firms, being mostly in the business of applying existing technology, may have greater immediate economic impact. As the firms develop, of course, it is possible that the academic links will prove increasingly useful.

Alternative strategies and conclusions

To summarize, we are looking at science park firms as diffusers and appliers, not leading-edge innovators. Such activities are important but cannot be confused with innovation. The growth of science parks has taken place at the same time as a steady decline in British science and technology funding (Chapter 1).

Science parks are filling some gaps, within the ideology of small, private innovators, and they are doing this using substantial proportions of state funding. Science parks represent large-scale public-sector investment: £57 million to date out of £92 million invested in land and premises. Local authorities have invested 23 per cent of the £92 million, or £21.2 million. Are there alternative perspectives for local initiatives, science parks and national initiatives, to promote science and technology for social and economic development?

British science parks have assumed a particular kind of US model. Models other than the 'private' US one may be more appropriate to Britain. We have examined the French technopoles (Chapter 7, p. 195), which are very large, linked to state decentralization policies, and completely state funded; and the Japanese recent attempt to use even 'science cities' to produce a breakthrough. Planned investment in these is $1 billion by 1991.

Such large programmes prompt the question, to what extent science parks are, or could be, linked to the bigger national programmes to stimulate technological development in Britain. The most recent new big programmes have been the Alvey Programme (Chapter 6, p. 169), British involvement in Eureka (Chapter 7, p. 193), Esprit (Chapter 3, p. 95 and Chapter 7, p. 192) and the Department of Trade and Industry's application programmes (Chapter 3, p. 89).

Large government laboratories like Harwell, National Physical, the Met Office, and so on (Chapter 2, p. 50), also still have large R&D budgets. There is no evidence that science parks have been drawn into discussion or involvement in these initiatives and organizations.

There must surely be opportunities to link local and regional industrial developments to the sites where major scientific and technological developments take place (e.g. the MOD, Chapter 2, p. 51; the UKAEA Harwell laboratories; etc.).

If national strategies for scientific and technological development should be more closely linked to programmes with potential use in production, of which Alvey is one variety, but not a particularly socially oriented one, it is important to co-ordinate smaller local networks, to ensure that regional and national initiatives do not contradict.

At local levels might it also be advantageous to site R&D engineers closer to production (as in Japan) rather than to site them in green zones away from existing industrial production? Science parks may act to separate R&D from production, rather than solving the problem of integrating technology and production more closely. Certainly the danger with science parks is that the property element takes priority over production and restructuring needs.

Outside the sunrise belts, which arguably do not in any event need science parks to attract high-technology industry, it is important both to recognize the need to be public-sector led, not to apologize for that, and to co-ordinate investment in high-technology industry with any property-led initiatives required. Surely within such co-ordination there is room for manoeuvre to include a role for socially useful production.

Any local initiatives to cluster innovative and high-technology groups should be planned within a coherent local and national strategy which could include increased local control of national science and technology. Public initiatives face a choice. They can either propagate the view that the solution to economic decline is small private entrepreneurship, or relate more closely to the local social and economic goals of elected councils and seek greater participation in development.

Wider higher education–industry interactions

There is now scarcely a higher education institution in Britain that has not sought to increase its association with industry.

In the public sector there is great commitment to vocationally oriented programmes, which calls for partnership with industry in their planning and implementation. Certain universities too run sandwich/integrated graduate programmes which enable students to spend a substantial amount of their time in industry. The value to industry of such co-operative planning and training is now well established. The students who have benefited from the programmes are more mature and have a greater awareness of the demands that industry will ultimately make on them. An additional benefit is that the staff members from industry and colleges learn to work together. The dialogue and regular social contact are the most fruitful source of joint research projects.

As already described (Chapter 3), a range of programmes have been drawn up to enable the partnership to flourish. It is, however, the initial personal contacts that are rate-determining. To this end, many universities have now established industrial professorships, which allow senior scientists from industry to spend a significant proportion of their time within the university. At student and faculty levels, therefore, greater association is being actively promoted. The number of academic professors who became directors of industrial companies is small but increasing.

The increasing proportion of industrial funds flowing into higher education is positive confirmation that joint collaborative research and contract research are being developed increasingly in university and college laboratories. Regular claims are being made that such work is leading to commercial products, but it is not possible at this time to quantify this contention. There are very few rich professors apparent at this time, who are enjoying the fruits of their exploited discoveries in Britain compared with the United States, where the trend has been in existence a great deal longer. Nevertheless, there are individual success stories. Surrey University, for example, has concluded a million-dollar deal with an Italian company for commercially developing a safe roofing alternative to asbestos. The Surrey Technology Centre is a recent addition to the university enterprises designed to trigger new, innovative high-technology companies.

The trend, therefore, is one of increasing commitment to utilizing higher education's human and physical resources to improve economic prosperity. It has been noted previously that these developments are proceeding against a background of great uncertainties with regard to higher education funding. Universities and public-sector higher education institutions will shortly have teaching and research funds identified independently. The objective is to establish strong centres of excellence operating over subject disciplines, even within a group of universities. Within universities too, the emphasis is on setting up broadly based research centres encompassing a group of departments. Inevitably, some universities or parts of universities will be excluded from these developments and be forced to spend more time teaching undergraduates, as occurs already within public-sector higher education institutions. Such strong centres will certainly be a more effective magnet for industrial sponsorship in the future. There is a danger, on the other hand, that basic research will cease to be a major thrust, except in a few highly selective centres. One thing is certain, universities will no longer be cosy institutions in which an individual could engage in his or her academic hobby. Targets, assessment, funding, and so on are the essential vocabulary for the future, and accountability for public funding will be an unyielding necessity.

European programmes too will exert a major influence on

successful British academic research centres. Five research programmes will share more than ECU 217 million (£143 million) if the European Commission's latest plans are accepted. SPRINT is a strategic programme for the innovation and transfer of technology. The principal phase (1989–93) proposes a budget of ECU 130 million. This will reinforce and extend the activities under way during the present experimental phase, which finishes at the end of 1988. SPRINT would concentrate on three main areas: strengthening the network of services and agents for technology transfer and innovation support throughout the community; support for pilot schemes in reception technologies, with priority going to underdeveloped community areas or those in industrial decline; and analysis of the innovation process to identify and remove obstacles and to formulate policy.

MONITOR is a three-pronged initiative into strategic analysis in the field of science and technology (SAST), forecasting and assessment in science and technology (FAST), and a support programme for a European assessment of research (SPEAR). The commission is seeking a total budget of ECU 22 million (£13 million) for 1988–92.

TELEMAN is a programme of research into remote handling in the nuclear industry, the aim of which would be to strengthen the scientific and engineering bases used for the design of nuclear remote handling equipment. TELEMAN would deal with basic problems associated with remote operations, using computer-assisted tele-operators; with questions relating to tele-operation in nuclear environments; and integration of the disciplines making up robotics. The commission estimates the cost of about ECU 40 million (£27 million), of which the EC contribution would be ECU 19 million (£13 million). Preference would go to projects featuring collaboration between industry, universities, and research organizations from more than one EC member state.

The fourth proposal involves a programme of decommissioning of nuclear installations. This would include nuclear power plants and nuclear fuel reprocessing plants and would continue work already under way.

Last, the commission proposes a programme of research in predictive medicine. This seeks to predict susceptibility to diseases with a view to their prevention and early diagnosis and improved prognosis and eventual treatment. There can now be no turning back. A university's success over the next few years will inevitably be linked with its ability to court industry.

Europe versus Japan

There is an enormous challenge, to both British industry and academia to harness the bewildering array of initiatives being launched from London and Brussels. Their declared aim is to counter the Japanese threat in science and technology. Meanwhile, Japan strives desperately to think and work in a more European manner. Might not a creativity transfer programme be launched from Europe and an associated creativity transfer programme initiated from Japan (p. 247)? Each side does certain things outstandingly well. This may well be the time for enhanced joint European–Japanese collaboration.

It could well be that the Japanese government is thinking along these lines, with the introduction of their Human Frontier Science Programme, which will, from 1989, be run by a specially created international foundation to be set up in Switzerland and funded from Japan. Fellowships, workshops, and research projects will be set up to support scientists from the European Community and the other Western industrial countries, Canada, the United States and, of course, Japan, provided at least three countries, including Japan, participate. The original announcement added that safeguards would be built in to allay fears that the programme is a Japanese ploy to increase their access to Western research. Why do the Japanese not insist on safeguards to prevent Western scientists from accessing the secrets of their highly successful application technologies ('associated creativity')? We would do well to enter into partnership in the belief that true co-operation benefits all participants; anything less is exploitation.

Appendix I

A cameo of Tsukuba Science City

This remarkable creation is a wonder to behold. On virgin land, unsuitable for rice production, near Tsuchura in Ibaraki Prefecture, 60 km north-east of Tokyo, a new modern city has been constructed since 1975 to promote high-level research in science and technology. The 'city' houses two universities, 31 government institutes, seven other government facilities, six government enterprise facilities, and seven private commercial research institutes. Most were transferred from Tokyo and rejuvenated in their new home. Total expenses for this creation amounted to ¥1,040,000 million to 1980. Homes and all other recreational facilities have been created anew for the present population of 130,000. This enormous project is in effect a shrewd investment of the profits resulting from the export boom of the 1970s. From every indication the move is already paying off handsomely. It shows what might have happened with British oil revenues.

Tsukuba's *raison d'être* is pure research and education, not R&D for commercial exploitation. In fact, manufacturing is completely banned at Tsukuba. The government hope is that it will be at the cutting edge in creating new technologies. Critics may argue that Tsukuba is an elaborate job creation scheme, biased in favour of the high-tech industries. However, it is a genuine attempt by the Japanese government to create new industries and technologies before its traditional industrial base is eroded, thereby forestalling the sort of industrial decline experienced in Britain. Even if Tsukuba has no fundamental effect on the Japanese economy as a whole, the regional economy in which it exists will benefit.

The University of Tsukuba is a new type of academic institution in Japan and is based on a system of clusters of colleges for student education. Research is organized around institutes and project research groups rather than single-discipline development. The style of management is also experimental, with boards of consultants appropriate to the particular activity drawn from the top industrial and academic community.

It is possible here only to list the main institutes and draw more attention to the MITI institutes:

National Laboratory for High-Energy Physics
Tskuba International Centre — designed to supply training for developing countries via the other institutes and in its own facilities
National Educational Centre Annex — to provide in-service training for schoolteachers
Tskuba International Agricultural Training Centre — for developing countries, mainly South-east Asia

A group of institutions specializes in problems of the construction industry:

National Centre for Disaster Prevention — problems related to natural phenomena, particularly earthquakes
Tskuba Telecommunication Construction Engineering Centre
Geographical Survey Institutes
Public Works Institute
Building Research Institute

The Science and Engineering group have some of the best brains in Japan:

National Institute for Research in Metals
National Institute for Research in Inorganic Metals
Tsukuba Space Centre
National Research Laboratory for Metrology
Mechanical Engineering Laboratory
National Chemical Laboratory for Industry
Fermentation Research Institute
Research Institute for Polymers and Textiles
Electrotechnical Laboratory
Industrial Products Research Institute
National Research Institute for Pollution and Resources
Meteorology Research Institute
Aerological Observatory

There are many more institutes dealing with agriculture, food, and health. These are clearly designed to be centres of excellence, providing mainly basic information that industry can have access to and exploit. There is the possibility also of drawing up contracts of a confidential nature to jointly work on commercial projects. MITI now is establishing more and more of its institutes at Tsukuba in a campus-type development.

Many scientists are mainly concerned with undertaking prestigious research that would bring them peer distinction by publishing in the leading journals and presenting their work at conferences around the world. MITI here is able to show off its best talents, in part as a counter to the hold that Mombusho has on this field in universities. The increasing tendency is to spend more and more of the effort on basic science and technology and to develop techniques. The approach is in tune with the country-wide desire to deliver its own basic inventions and breakthroughs. A similar situation existed in the United States some ten years ago, but has now retrenched by concentrating more on predictable developments. A significant new feature was the requirement that all scientists should undertake their own measurements. The best equipment is provided, mainly from Japan, some from the United States, and occasional items from the Soviet Union. No technicians run about the place, since the scientists themselves are forced to serve this role also, even to developing their own X-ray film and typing their papers.

Here is an excellent way to prevent scientists becoming deskbound, as happens more often than not to senior scientists in Britain. Graduates, therefore, are the doers and innovators, trading off instruments and techniques from one another. There is, therefore, much collaboration and material discussions of projects, albeit a sense of emptiness in the absence of support staff.

To illustrate the workings within Tsukuba, we can consider the following project examples. Polyvinyl alcohol was converted to a hydrogel with gamma radiation and fabricated so that it could be used as replacement fluid in the vitreous body of the eye, for example, to treat detached retina conditions. To develop the project from basic production to acceptance by the health regulatory authorities, chemists, biologists, and physicians had collaborated closely. No one had deliberately drawn together an interdisciplinary group to deal with the project, as often occurs in British and US companies, but sheer necessity had forced the originator of the work to call on successive experts to complete various facets of the project. They were enlisted, therefore, at the exact point where their expertise was called for. The success engendered a collective pride. Together, the group recognized other bio-engineering uses for the new material, and planned its exploitation as a drug delivery system. It should be noted that the university medical school was also involved, illustrating the ready university–research (industrial) institute collaboration that is possible in the Tsukuba set-up. There is a freshness and stimulating atmosphere, which must do wonders for the productivity of the academic community, which in effect is what the MITI institutes are.

Another peculiar Japanese element is the rather special way decisions are arrived at. These emerge, rather than being issued from one or a collection of powerful individuals. The bench researchers, with their colleagues, formulate the proposed project, which is then discussed at any one of up to nine peer groups. The consensus goes to MITI's Agency of Industrial Science and Technology. Then in mysterious and unknown ways, groups from among the 2,500 scientists employed in administrative, evaluatory, or planning capacities by MITI chew over the projects again before they return with modifications, approval, or rejection (very infrequently) to the operative institutes. As one researcher said, 'If we do not know even now who was responsible for starting the Second World War, we cannot possibly know how major decisions are taken at MITI.' No one individual or even one single group knows. Tsukuba science city has no counterpart anywhere in the world; it is unique.

Appendix II

Co-operation of industry in Japan with higher education

Makoto Kikuchi
Managing Director, Sony Research Center, Yokohama, Japan

In the first part of this appendix, I will give a personal view of the past 35 years of R&D in the Japanese electronics industry. I will describe the relatively fast catch-up of Japanese industry in this field after World War II, and will give possible important factors closely related to this rapid progress. At the same time, I wish to point out the change in the nature of technology in these 35 years, which resulted from a 'maturing process'.

In the second part, I will discuss the contribution of university professors to the Japanese electronics industry.

And in the final part I will discuss the problems we have in Japan on the future co-operation of industry and higher education concentrating on the next 10 to 20 years. The problems are related to two factors, namely (1) the position of Japan internationally and (2) the expected contribution of basic research to the next breakthrough.

Background

I graduated from university in 1948 and joined a government laboratory belonging to the Ministry of International Trade and Industry, where I started my research on the physics of semiconductors. This was a happy coincidence for me, because the first announcement of the birth of the transistor was also made in June 1948 by Bell Telephone Laboratories. We (the Japanese people) were still very poor then, and suffering from lack of food and many other necessities. The trains were extremely crowded and I remember standing on the carriage connectors in the morning and evening rush hours.

In the early days of the research on semiconductors and transistors in Japan, very few scientists could understand why germanium crystals revealed amplification when two fine needles were set on the surface of the crystal. We learned about the science of semiconductors through scientific journals mostly from the United States, such as *Bell System Technical Journal*, *RCA Review*, and the *Journal of Applied Physics*. We were extremely excited to do research on this unknown material, and worked very hard.

The government laboratory we worked in was just a simple wooden building and the temperature inside was almost the same as that outside. We had no central heating and of course there were no air conditioning systems. That was the condition of our laboratory in the late 1940s.

In 1955 the transistor was produced by a Japanese company, and the first transistor radio was introduced into the market by Sony in 1956. The Japanese transistor industry started to grow from the middle of the 1950s.

My first stay at the Massachussetts Institute of Technology started in 1960, working in the Electronics Research Laboratory under Professor Jerome Wiesner, who had recently left Cambridge. I joined the International Conference on Solid State Devices in 1963 at Michigan University, where I tried my first formal presentation in English. I have spoken at many more conferences and seminars since then in more than 15 countries, and I have made many friends throughout the world in these past 25 years.

In 1974 I resigned from MITI's laboratory and moved to Sony and started to learn if a physicist could be useful in the battlefield of the electronics industry. My job changed from that of an independent research scientist to a manager of an industrial R&D establishment.

Moving from my personal background briefly mentioned above, I would now like to discuss some of the problems underlying the relationship between industry and higher education in Japan.

The fast 'catch-up' of the Japanese electronics industry

Before examining the industry–university interaction, I think it is necessary to consider two historical views. The first one is shown in Figure A2.1. This is my personal view of the Japanese electronics industry over the past 35 years.

The black circles represent Japan and the grey circles represent Europe and the United States. This mountain climbing analogy shows two things: the reduction of the distance between Japan and the advanced countries and the retardation of the advanced countries by the increase of the gradient. This is related to the historical change in the nature of technology caused by the maturing process.

Looking back over the history of industry, I am frequently

Figure A2.1 Mountain climbing — a model of the Japanese electronics industry

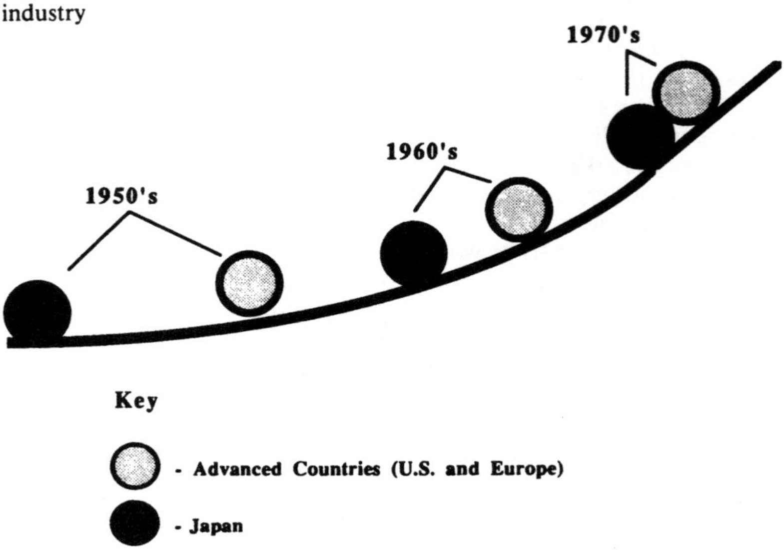

reminded of the expression 'Made in Japan'. One day in 1947 I was in a theatre watching an American movie. Bob Hope was acting as a pirate and had an iron hook on his left arm. He was fighting against other pirates on a boat and finally his iron hook was broken just by a blow of a sword. He checked the broken hook and said with a sneer on his face, 'Ah, it's made in Japan'. I clearly remember that this expression was widely accepted during and after World War II, meaning 'reasonable price but poor quality'. That was synonymous with Japanese products for a long time.

Four factors supporting the fast 'catch-up'

I am frequently requested these days to give talks at seminars and symposia in the United States and Europe on R&D activities in Japan. I usually start my talk with Figure A2.1, because I know that our position in the world was very low in those early days. In most of these seminars I have been asked my opinion on the principal factors that supported the catching-up process of Japanese industry after the serious damage of the war.

My response is summarized in Table A2.1. I think these four items are important, and they are related to Japan's more recent problems, such as the imbalance of international trade.

Table A2.1 Key factors involved in Japan's fast catch-up with the United States and Europe

1. Enthusiasm
2. Homogeneity (education, etc.)
3. Second runner
4. Maturity of electronics

Source: Compiled by the author.

Enthusiasm is an important factor, because Japanese industry grew mostly by the people's drive and not by a government push. My American friend, who was a professor at Harvard University, once talked to me about his simple impression of Japanese students. He pointed out the general behaviour of Japanese students he observed in the Harvard Square, that they were always walking very fast compared with other students. They looked like they were always making haste.

Another friend of mine told me of a more ironic example. He said, 'In the boarding bridge at the airport which connects the building and the airplane, if you find someone passing other passengers in it, that is a Japanese.'

Generally speaking, we Japanese always try to accomplish things in a short time, which is very different from the behaviour of our Chinese friends.

I graduated from university three years after the end of World War II. We were suffering from lack of food and very bad social facilities, but I remember I was extremely enthusiastic about my research on semiconductor physics and transistor science. I do not mean that enthusiasm is always good, but I am just thinking of the historical and cultural situation. Our enthusiasm for industry sometimes results in a factor that makes Europeans see us as 'a noisy and busy nation'. At any rate, this factor was important in the catching-up process.

The second factor, homogeneity, has several meanings. The point I wish to emphasize is the homogeneity of education, especially primary and middle school education, which has been standardized in Japan. This education pattern was initiated about 100 years ago, in the Meiji era, by the national leaders of Japan. The leaders thought that the most important target of the Japanese society should be to catch up to 'western civilization'. They thought that to achieve this target it would be necessary to implement a policy of education for all the people in Japan. In the last 100 years education has become more and more widely accepted in our society, and more and more people have been trying to reach the higher levels of education.

Table A2.2 shows the percentage of students going from high school to universities. As shown, almost 94 per cent of the students from middle schools advance to high schools, and about 30 per cent of them go to universities. I am not talking about the 'quality' of students but merely the popularity of education in Japan. Most Japanese parents are extremely eager to send their children to a 'good university'. To achieve this they must choose a 'good kindergarten'. Many Japanese feel this situation is undesirable because they know that the competition for the entrance examination is now becoming too intense, but they cannot easily escape this situation. This stems from our historical and cultural background, which is not easily changed.

Table A2.2 Percentage of students going from middle school to high school and from high school to university in Japan

Year	*To high school (%)*	*To university (%)*
1970	82.1	24.2
1975	91.9	34.2
1980	94.2	31.9
1984	94.1	29.6

Source: Compiled by the author.

However, the homogeneity of education, even to high school level, was important in bringing the level of Japanese industry up to that of the United States and Europe.

The third factor, termed the 'second runner', is due to historical facts. Japan jumped into modern industry after World War II, as shown in Figure A2.1. That means that there were mountain climbers in front of us already. The first climbers' group was there and we were the follower. It is important to note that this situation was not our choice, but just the historical reality.

Now, the first climbers were called 'pioneers', and pioneers have always had to pay a price to be pioneers. This price was the effort to try new exploratory research work. The following is a typical example. In the early 1960s there was an interesting topic called 'electronic cooling' or 'Peltier cooling'. The effect could be observed just by passing electric current through a junction between two semi-conductors. By proper choice of the direction of the current flow, the junction was cooled. This topic was of great interest to pioneers then. The United States spent almost $40 million on it in three and a half years of R&D. Finally, the pioneers came to the conclusion that nothing would come of it and this was really a loss for them.

Japan, as the second climber, could observe the course of events in those days, and could learn much from the pioneers. We were, of course, doing our own research on this phenomenon in our own laboratories, but before spending tremendous amounts of money we could learn from the experience of others. In this way, Japan could save money, manpower, and time. I believe this factor was very important in the catching-up phase of Japanese industry.

I will not mention the fourth factor here, but will touch upon it later.

Merits

Japan is now considered to be industrially advanced. Frankly speaking, however, I have noticed a tendency to over-rate Japanese ways of doing things by my foreign friends, due mostly to a distorted image of Japanese people and society.

A typical example is as follows. Several years ago, when I was responding to an interview for a British magazine, I found that the interviewer had an incorrect image of Japanese government, especially the Ministry of International Trade and Industry (MITI). He thought that the bureaucrats of MITI were like supermen who could understand clearly the right direction for Japanese companies. For him, every Japanese industrial success stemmed from the guidance of MITI. I tried very hard to correct this misunderstanding, but his belief was unshakeable. Finally, he asked me a question.

'Now, Dr Kikuchi, please answer my question frankly.'

'Yes, I would like to, of course.'

'Now, how many times in a year do MITI bureaucrats call up the president of Sony, and tell him to do such-and-such research and development and offer a financial support?'

This question surprised me very much so I responded, 'I have never heard of such a case in my life.'

'Oh, you still wish to hide it.' He shook his head.

When the BBC came to my office to record an interview for a radio programme, I explained about my mental picture shown in Figure A2.1. The interviewer asked me, 'Where is Japan in the 1980s?' Is Japan the real pioneer now, or still a follower?'

I told him that Japan has been doing rather well in certain fields, while in other fields we are still lagging. Fortunately, the average level of Japan's industry is not so different from that of western countries, but it is very important to recognize the situation in

Table A2.3 Areas of technology in which Japanese and western industries are performing well

Japanese industry	*US and European industry*
D-Ram (256K) production	Microprocessors (8 bits, 16 bits, 32 bits)
S1, Galium-Arsenide (GaAs)	Software
Crystals	Photoresistors
Packaging	Electron beam
	X-ray
2 μ rule	1.2–1.5 μ rules
Automation	X-ray litho
Junction lasers	Computer-aided design
CCD	Military
	Special uses

Source: Compiled by the author.

individual fields. Table A2.3 shows some examples of the fields of technology in which Japan and the other advanced countries differ.

The left side shows the fields in which Japanese industry is performing well, and the right side gives examples of the fields in which the United States or some European countries are much more dominant. For example, production of very large scale integration (VLSI) is now the area in which Japan is doing well. Also crystals used for the semiconductor industry are well provided for in Japan. On the other hand, in next-generation microprocessors, the United States is ahead, and software research and development is an area in which Japan is definitely lagging. New types of high-precision processes, such as molecular vapor epitaxy, are mostly designed and modelled by US or European companies, and Germany is dominant in the field of high-vacuum instruments.

The list of fields shown in Table A2.3 can be summarized in the way shown in Table A2.4. Here I wish to classify the concept of creativity into two types.

Table A2.4 Different types of creativity shown in the west as compared with Japan

Japanese (associative creativity)	*Western (independent creativity)*
Target-oriented	Target-finding
Action	Idea
Co-operative	Personal
Modification	
Improvement	
Systematize	

Source: Compiled by the author.

The original meaning of the word 'creativity' can be described as independent creativity, which is essentially a very personal activity that opens up an entirely new field. This approach can be called 'target finding'. There are many examples of this type of creativity in the history of science, such as Newton, Michael Faraday, Einstein, and coming down to the recent examples of Bardeen and Shockley. On the other hand, I think we have another type, called 'associative creativity'. This approach is not really target-finding, but it is target-oriented. In other words, it starts with some hints. Although it depends on the hints given, its approach is also unique and creative. The transistor was invented by three scientists, Bardeen, Shockley, and Brattain, but there has been another type of contribution in the electronics industry that tried very hard to introduce the transistor into consumer products. An example was the world's smallest radio. Another effort was to put transistors into television. I think these types of contribution have been unique and very valuable for society, and should not be ruled out from the concept of creativity.

The term 'associative creativity' comes from associative memory, which relates new concepts with a given idea. This type of creativity is, in its nature, effective in modifying and improving existing ideas, and in systematizing scattered ideas. It is also based essentially on good co-operation.

Cultural background

Let me sum up the core problems here from the things mentioned in the preceding sections in connection with Table A2.4.

I have looked back over the past history of 35 years and shown certain technological fields in which either Japan or the advanced western countries have been doing better. By the classification given in Table A2.4, the characteristic contributions to the historical development of science and technology from different sides can be seen. In short, to open new fields, western countries showed definite strength, whereas Japan has revealed strength in target-oriented activities. The former led to invention and revolution, the latter contributed innovation and evolution.

This view seems to be correct as far as the history of the past 35 years is concerned. Invention of the transistor, invention of the junction laser, information theory, and so on, are examples of entirely new fields that originated from western creativity. The progress of VLSI, reliable production of LSI's commercially available junction lasers, and many consumer electronic devices and systems such as the compact disk and home video recorders are examples of associative creativity.

The Japanese approach, characterized by associative creativity, has its roots in its cultural and social background. In Table A2.5 a cause-and-effect relationship is given. The strong motivation I mentioned earlier is caused by the enthusiasm of parents for good schools. Why a good school? One of the most important reasons for this is the element of competition in Japanese society. Graduation from a good university is essential for getting a good job, which in turn is necessary for survival. Why such competition? One of the important reasons is the high population in Japan. I have no doubt that the Japanese culture would have been entirely different if we were living in the vast territory of the Soviet Union. The educational pattern and resulting homogeneity of education, with the unfavourable factors of being geared towards entrance examinations, gives rise to a feature shown in Table A2.5, namely high learnability, hint-oriented creativity, and a capability for innovation. Innovation is essentially different from the invention. Invention is the result of independent creativity.

Table A2.5 The cultural and social background to Japanese innovation

Educational background (primary, middle)
from
'Good-school' orientation
from
Competition of people
from
High population on small islands
leads to
Learnability
Hint-oriented creativity
Innovative attitude

Source: Compiled by the author.

Historical trends in electronics technology and industry

Before coming to the main part of my discussion, I think it is extremely important to survey the historical situation of modern industry and related technologies. Figure A2.2 shows my mental picture of recent history. Time flows from the top to the bottom in this figure. One of the two lines shown there, which starts with the vacuum tube (VT) shows the flow of hardware. The other line, starting with operations research (OR), shows the flow of software.

Figure A2.2 The interaction between vacuum tube (VT) technology and operational research (OR)

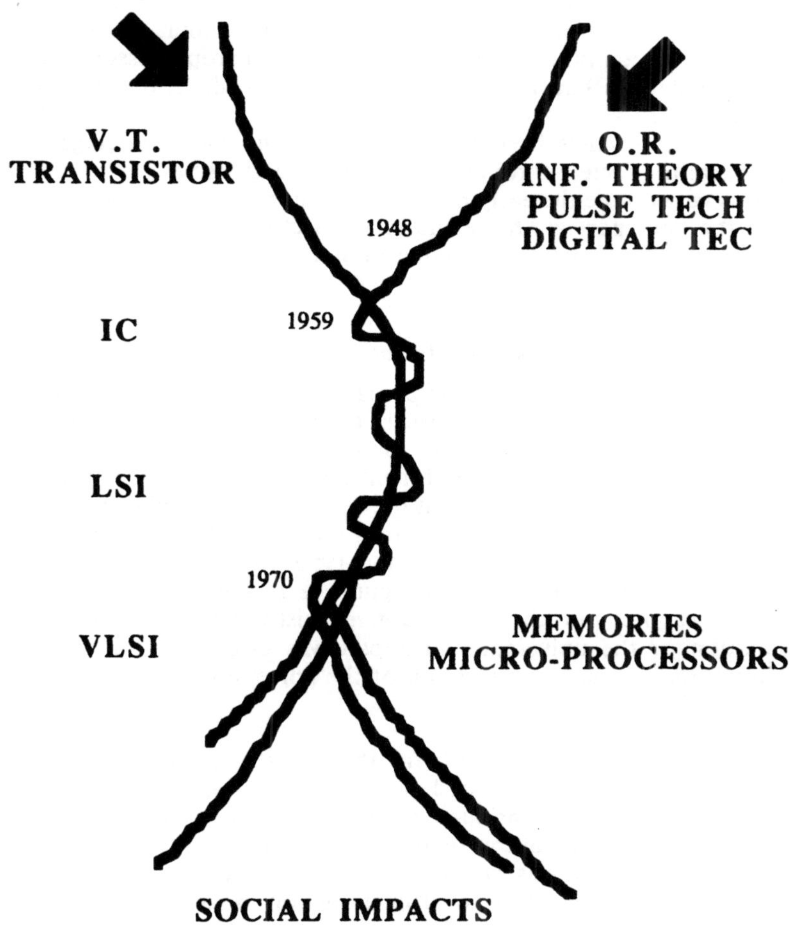

An important aspect of this picture is the point of mixing of the two flow lines around 1959. The point is that because of the revolution caused by the birth of the transistor, the integrated circuit (IC) was born from it in 1959. These two events made it possible to combine the flow of hardware and software, which had been independent and parallel before that time. The combination of the two flows accelerated the progress of integrated circuit technology, especially microprocessors, which gave strong stimulation to the progress of computers. From around 1970 microprocessors and relevant devices had a tremendous impact on all fields of technology. It also had more and more influence on society and life styles. Industry has been providing more and more 'intelligent' products for society through the use of IC and VLSI.

I usually define the early part of this history as the 'revolution', and the later part as the 'evolution'. Electronics technology has shown a maturing process just like a human being, showing growth from a baby to an adult.

Contribution of university professors

Now, on the basis of the ground prepared by the discussion given in preceding sections, I wish to discuss the relation between industry and higher education, especially Japanese university professors.

From the historical analyses I mentioned in the early part of this paper, I think it is now clear that the contribution of universities to Japanese industry was different from that in the United States or European countries. In the 1940s and 1950s Japan was the second mountain climber, as shown in Figure A2.1. Again, I should say that it was not our choice but a given historical condition. The situation of the universities in Japan was essentially the same. Therefore, the main contribution of the university professors was as follows:

1. The most important task for them was to provide industry with reasonably well educated students. This was achieved with relative ease, as I mentioned earlier; both the professors and students learn easily and were able to catch up quickly.
2. The other task was rather special. I must mention here something about the nature of government action in Japan. MITI has made many contributions to Japanese industry, but things are not exactly as my American or European friends believe.

Believe it or not, the bureaucrats of MITI are almost laymen of science and technology. Powerful MITI bureaucrats are innocent

Figure A2.3 An example of government aid to a project

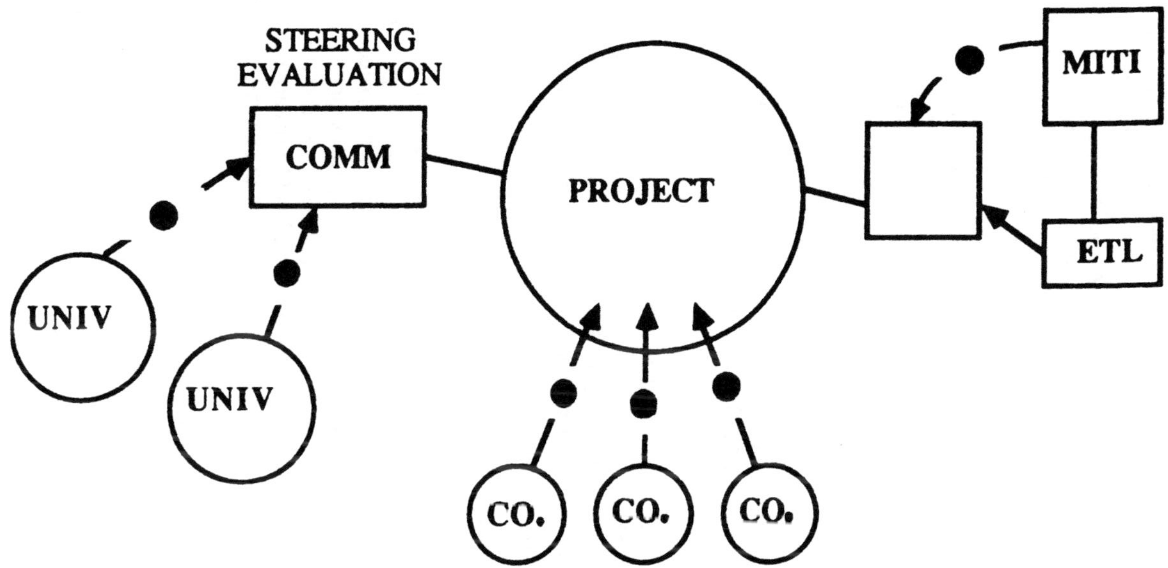

about details of technology. Because of this, they never hesitate to ask simple question on modern topics such as electronics. I often received telephone calls from my MITI friends and they asked me, 'Tell me about GaAs. What are the merits?' or 'Where is VLSI going in the next 10 years?' The answers to these simple questions are extremely important to managers and supervisors. These types of questions are much more important than those of detailed science.

Now MITI has long been dependent on university professors in order to get a clear view of modern industrial technology. In other words, this has been an important contribution of university professors. Japanese professors were sometimes looked down on by American professors as 'encyclopedic', as just a substitute for a textbook, working only for learning and storing instead of creating knowledge. However, helping MITI bureaucrats to judge and make plans for Japanese industry, professors have made many contributions. In Figure A2.3 the structure of government aid in Japan is illustrated.

When organizing a large project, MITI usually assists indirectly. In the working group that is to do the research and development, staff members are sent from the companies that applied for the grant from MITI. As shown on the left side of the figure, there is a committee for steering and evaluation. In this committee, university professors usually join and give advice to MITI bureaucrats. They also serve to evaluate the quality of the work being performed by the members sent from companies and advise on the allocation of the budget.

In this way MITI bureaucrats have always been willing to ask questions of professors on modern technology, and to seek assistance in the management of technological projects. Because of this relationship and because of the professors' 'encyclopedic' knowledge, MITI seldom made serious mistakes. This situation might have given the impression that MITI bureaucrats were supermen who could always guide Japanese industry rather successfully.

Co-operation between the university and industry

As already mentioned, Japan joined the electronics industry as a follower in 1948, and electronics technology has been changing its nature from revolution to evolution in the last 15 years.

In the past, there has been very little direct co-operation between universities and industry in the electronics field. However, I can tell you of some rare cases of co-operation.

Sony once had a very nice contract with a professor of Tohoku

University. In this case the professor had been working with Sony for many years before, so he had a friendly relationship with many Sony engineers. The most important point, however, was that his idea on 'perpendicular magnetic recording' was unique. I believe this is really the most important factor for considering co-operation between universities and industry. We recognized that his idea would be very attractive for our purposes, so we sent some engineers to his laboratory and let them join his group, and Sony gave him reasonable financial support.

After several years Sony skimmed the cream from his idea, and came up with its own way to design a practical recording head. Then I told him that we would proceed on the basis of what we had learned from his original idea. That was a friendly negotiation, and Sony disclosed its own design of a 3.5-inch floppy disk using perpendicular magnetic recording. This, personally, was really an enjoyable example of the co-operation between university and industry. Again, I emphasize that the most important point was the uniqueness of the idea.

Frankly speaking, I think we should try to improve the present situation of co-operation, because Japanese industry is gradually joining the first climbers' group. We must try to find a nice 'seed' idea in a university and take it through to the development and production level. There are several problems, however.

From the point of view of the university professors, the problem is to know industry's needs. When I was working at the Massachusetts Institute of Technology in the United States in the early 1960s, my friend Professor Dick Adler spent his summer vacations at Bell Telephone Laboratories. This not only compensated a rather lower salary, but he also could experience 'industrial technology' from the inside. This type of opportunity is, however, very difficult for Japanese professors to find because of the conditions of our society.

The difficulty comes, for example, from the serious competition between companies in Japan. There are many 'me too' animals in our society, so every company is exposed to severe competition. Therefore, it is almost impossible to keep university professors even in the laboratory, because the professors have no sense of keeping secrets.

This lack of feeling for industry causes another undesirable effect on the students. Once I found a student who believed that his career as a researcher would be seriously jeoparized if he joined a company. I tried to talk with him again and again. I finally found the reason for this. His professor was always telling his students that his research was the world's best, and no company was doing equivalent research. That professor did not know that a company laboratory would perform fundamental research if that was necessary for accomplishing the target of new development.

University professors have another problem these days, because of the maturation of technology (semiconductor technology, for example is now instrument-oriented). In other words, to do reasonably meaningful research, we need instruments that are very expensive. We have many universities now, and we have many professors in Japan. Financial support by the Ministry of Education is limited, so professors are always suffering from a lack of money to research interesting problems. MITI does sometimes offer professors the opportunity to obtain money, but this causes another problem affecting the independence of universities.

Really important problems

Figure A2.4 shows again the historical situation of the electronics industry, starting from the invention of the transistor in 1947 through to the integrated circuit. From the technological revolution we are now in the era of 'steady evolution'.

As mentioned earlier, inventions arose mostly from independent creativity in western countries. Japan, a follower in early days, gradually grew to contribute to the evolution through 'associative creativity'. These contributions were supported by the characteristic merits of western and Japanese types of approaches, which were related to cultural and social backgrounds.

Now we have two tasks in front of us. One is the continuing evolution, and the other is to prepare for the next possible breakthrough. In short, the western type of contribution was performed by outstanding individuals; the contributions from Japan resulted mainly from people with homogeneous educational backgrounds. Many innovations in consumer electronics had been performed in this style. From this it can be concluded that for the continuation of steady evolution, Japan is in a better position because the nature of the work in supporting innovation is conducive to the Japanese style. However, the inventive type of work should be fitting to the western approach.

Japan needs outstanding individuals corresponding to those of western countries who made outstanding contributions in the history of science and technology. In short, the problem is, can Japan produce true inventors in the near future? I have made an observation that gives me optimism. Figure A2.5 shows the statistical results of my research. I examined many good scientific

Figure A2.4 A history of the electronics industry

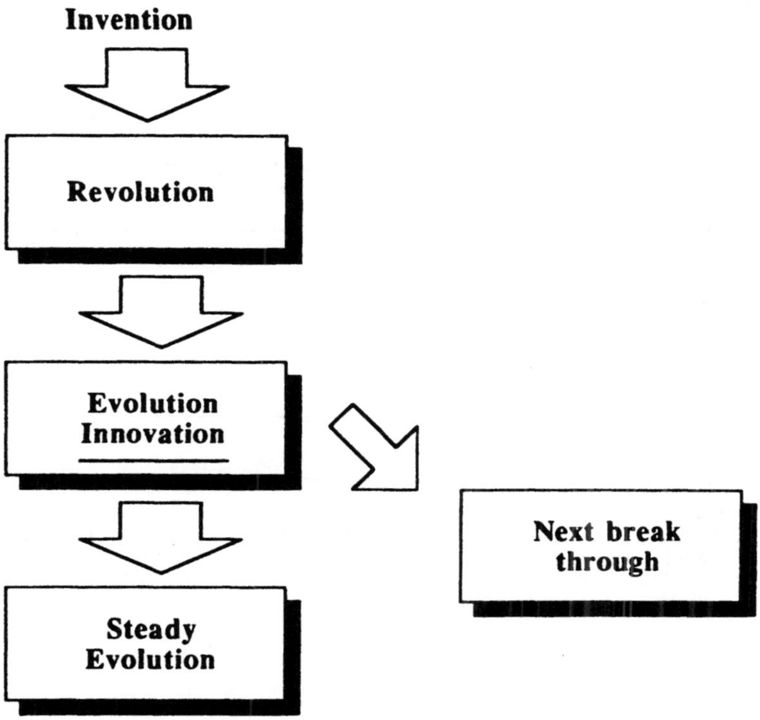

Figure A2.5 Japanese references in the *Journal of Applied Physics*, (GaAs — Lasers; 1962–82)

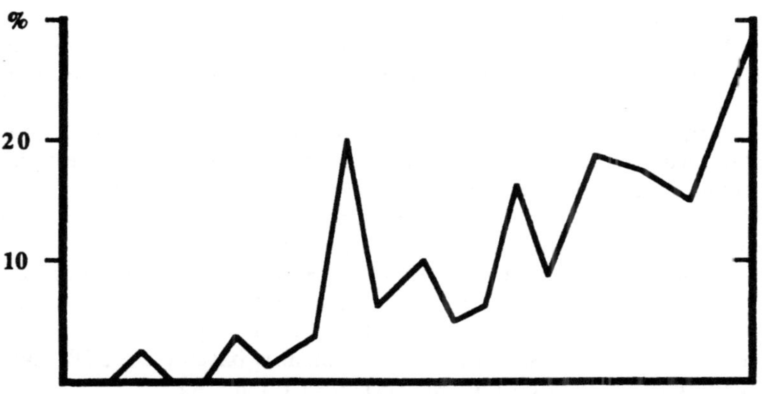

Figure A2.6 Grants-in-aid for scientific research (trend in budget)

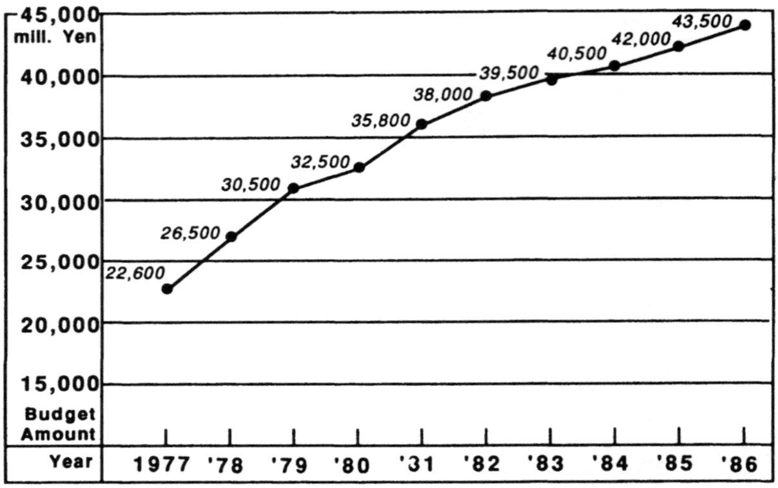

papers by American or European authors in international journals in several fields of research.

The results on the *Journal of Applied Physics* are shown in Figure A2.5. I counted the number of papers by Japanese authors in the list of references at the end of those papers. Starting from 2 or 3 per cent in 1964, it has increased to 23 per cent in 1982. This statistical result clearly shows that the quality of papers by Japanese authors has slowly but ever increasingly improved over the past 20 years, because citing a paper in the reference list means that the author attaches scientific importance to that paper.

I remember when I gave a talk to the Royal Institution in London for the Friday Discourse in 1985, I closed my talk by saying, 'for the tomorrow of Japanese electronics, I have 60 per cent optimism and 40 per cent pessimism'. My optimism is based on the 20 years of increasing research budgets shown in Figure A2.6.

Conclusion

Japan has made considerable progress in catching up with western advanced countries. The Japanese contribution was mainly the innovation type, whereas new fields were opened by outstanding westerners.

We are now facing two tasks. One is continued evolution, and the other is to prepare for the next breakthrough, which are quite different from each other. We must try long-range efforts to produce individuals who can create new openings in science and technology. Our fundamental social and historical factors have led to homogeneous education, and it will be a difficult task for us to change this intellectual pattern in a short time. But it is one of the most pressing challenges of our time. The most important tasks for the university and professors are, I believe, related to the problems raised in this paper.

Appendix III

Co-operation of higher education in Japan with industry

Professor Masamitsu Ohta
President, Nagoya Institute of Technology

Higher education in Japan is supported by 466 institutions; 124 (including 53 national institutions) of these are engaged in the teaching of science and technology, and the annual enrolment is over 80,000 students.

These figures are a result of the efforts during the past quarter of a century of improving the teaching of science and technology in Japan. Such advancement has enabled industry to obtain a good supply of engineers for the technological development and delivery of excellent products, resulting in the increasing prosperity of the country today.

The remarkable growth of Japanese industry may well be attributed to the nature of the Japanese people and social structure of Japan. The Japanese, living in their narrow islands as an agricultural people throughout their long history, are characterized by immobility and a conciliatory attitude, which has brought forth the whole-life employment system and zealous group activities in organizations.

Technological development and production can, by nature, advance most effectively in such a social atmosphere, and the remarkable progress of Japanese industry may well be attributed to these characteristics. Meanwhile, the improvement and expansion of higher education in the post–World War II period dramatically increased the number of educated people and caused the change from a hierarchical society to a society dominated by the middle class. This change of social structure has been conducive to the recent evolution of industry. Technological development has caused a change in the labour force requirements of quality and quantity, resulting in a fast-growing knowledge-intensive industry that needs a large number of intellectual workers. The technological colleges, in such an environment, have a high social profile and are run under progressive management. However, it cannot be denied that there are still many problems in higher education, in particular in developing effective co-operation with industry. In this paper, university–industry co-operation will be discussed from the viewpoint of the university.

Research activities in universities

The primary function of a university is to promote the development of science, and research activities in universities are centred around basic research. However, applied research has a particular importance in technological colleges. This is because the development of a new technology originates in basic research and is completed when it is put into practical application after applied research (see Figure A3.1). The fundamental pattern of research in universities is to start with basic research and to shift the research subject to the application stage. As for the personal history of a researcher, it is desirable to devote oneself to a basic research subject when young and, after acquiring the basic capabilities required of a researcher, to expand the research activities to the area of application.

One of the features of research in universities is the freedom of choosing the research subject; a university researcher can select the subject at will. This means that research in the university is basically carried out individually, with the help of assistants and graduate students. However, two or more researchers often form a project team, with each researcher responsible for a part of the work.

The project team is not confined within a university. Joint research activities are carried out in various manners — in co-operation with researchers in other universities, research institutes, or private corporations.

Researchers in universities are provided with a scarce supply of funds. In the national universities in Japan, the ordinary annual budget is divided among the researchers. But the amount of each share is so small that research facilities can only be purchased with the aid of other sources of research funds. The outside funds are, however, limited in number and researchers have difficulty in obtaining funds for facilities.

Thus many researchers in Japanese universities seek sources outside the ordinary budget of the universities, such as grant-in aid for scientific research, academic research promotion funds, joint research funds, contract research funds, and donations.

The grant-in-aid for scientific research is a fund granted by the state government, and the amount in fiscal year 1986 was ¥43 billion and was granted to 16,000 research subjects. Although the amount of this fund is increased every year, the number of

Figure A3.1 Progress of research and development

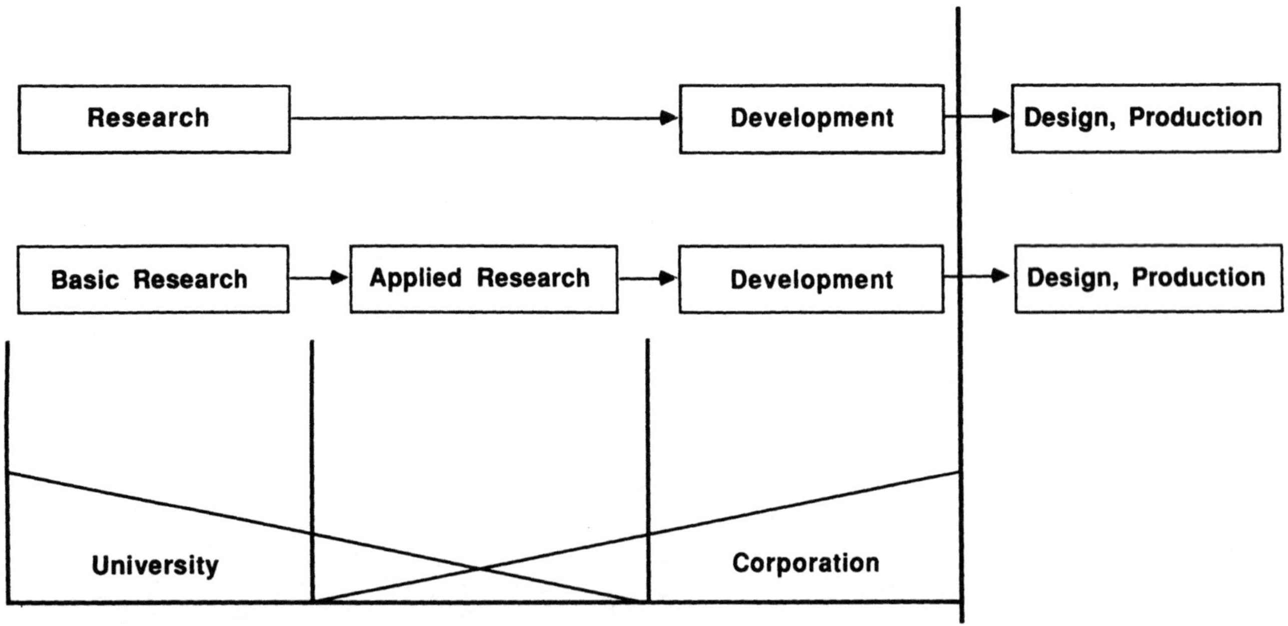

Figure A3.2 Grants-in aid for scientific research (trend in applications and approvals)

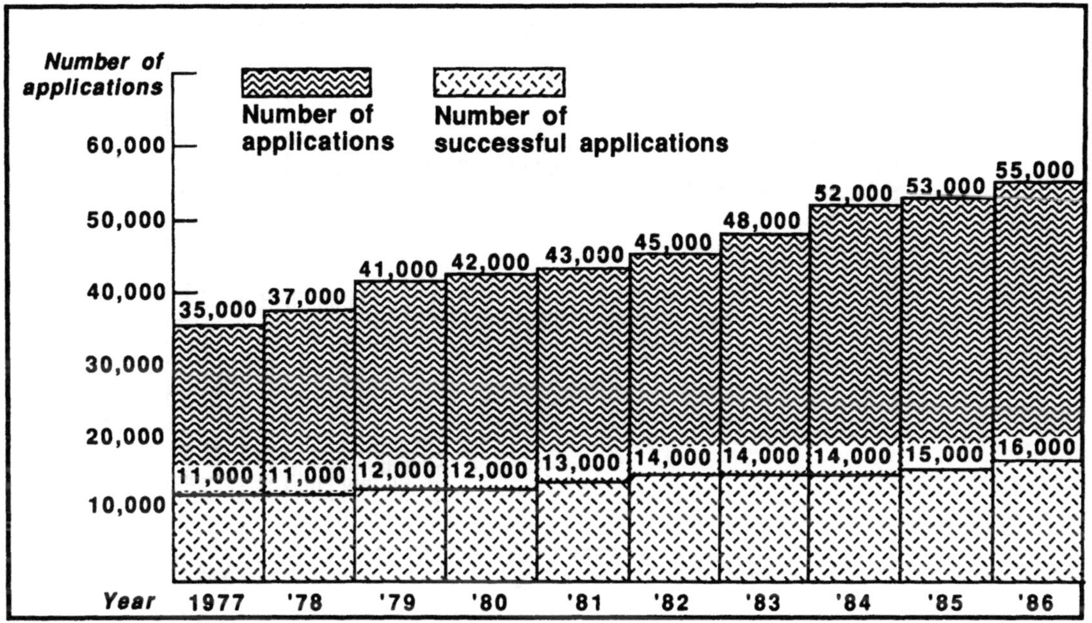

Figure A3.3 Number of joint research projects and total budgets for the projects supplied by the Ministry of Education, Science, and Culture

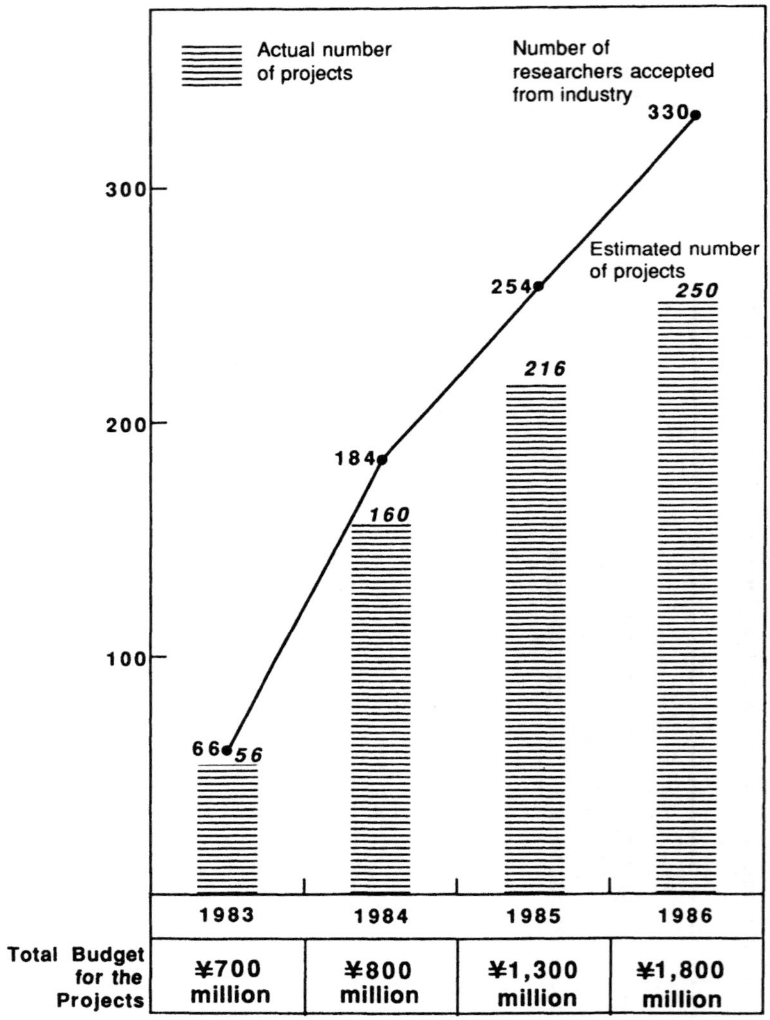

applications is so large that only 30 per cent of applicants are successful (see Figures A3.2 and A3.3).

The academic research promotion funds are sponsored by private corporations, and are granted in wide areas of research, but the amount of fund per research subject is inadequate.

The last three sources of funds mentioned above are those provided for particular research subjects based on co-operation between the university and industry, and will be described later.

Researchers in universities must utilize these valuable funds as effectively as possible. Co-operative utilization of research facilities is an important measure for this purpose, and each university is making efforts to utilize, maintain, and manage the research facilities effectively by, for instance, establishing research facility centres. On the other hand, universities are provided with an ample supply of research manpower. A professor normally has one or two assistant graduate students and undergraduate students under his or her instruction.

The management functions to promote research activities in universities have also been improved. In our institute, research co-ordination officers have been established this year to enhance administrative functions, and an Academic Research Promotion Committee has just been established to promote project research.

Present state of university–industry co-operation

Universities have the facilities for scientific research and industry has the facilities for production. Thus university–industry co-operation is the joint research of science and production aimed at the development of new technology through the interaction of these two factors. Actual forms of university–industry co-operation can be classified into the following two categories. The first is the case where basic research in the university has achieved considerable progress with innovative invention and analytical works, and industry tries to put these accomplishments into practical application. In such a case, as there remains much to be studied following the basic research, the research in the application stage is to be done jointly by the university and industry.

The second is the case where a company has accumulated technological know-how in an advanced industrial field, and it can utilize this in the next stage of development. In such a case researchers from universities and private corporations join forces. The significance of this lies in the close co-operation of the researchers from both parties, and in that creative work can result from the exchange of knowledge and discussion. In such a joint research project, the

Figure A3.4 Trends in the donations received by national universities (1975–85)

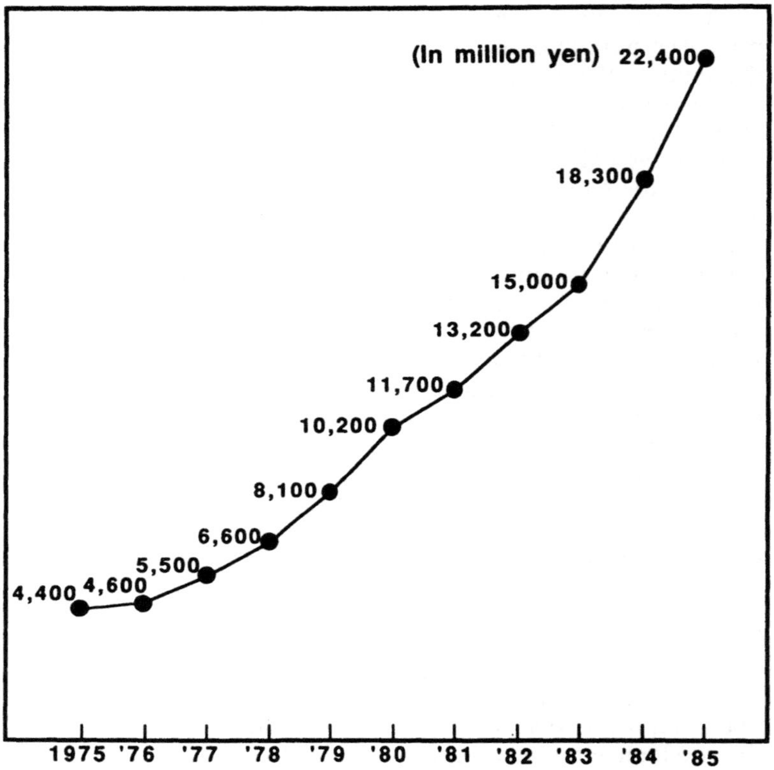

university can absorb the state of designing and manufacturing technologies and compare this to its own basic research, and the industry has the advantage of acquiring scientific background. The latter form of joint research is predominant in university–industry co-operation today.

The Japanese government has established a new policy to promote joint research by national institutions and private corporations. This has led to closer co-operation since its conception (see Figure A3.4). The research projects in which our institute is involved this year are shown in Table A3.1.

In Japan the relationship between universities and industry is similar to that which exists between universities and government institutions, government institutions and industry, and the user

Figure A3.5 Contract research projects and funds received by universities

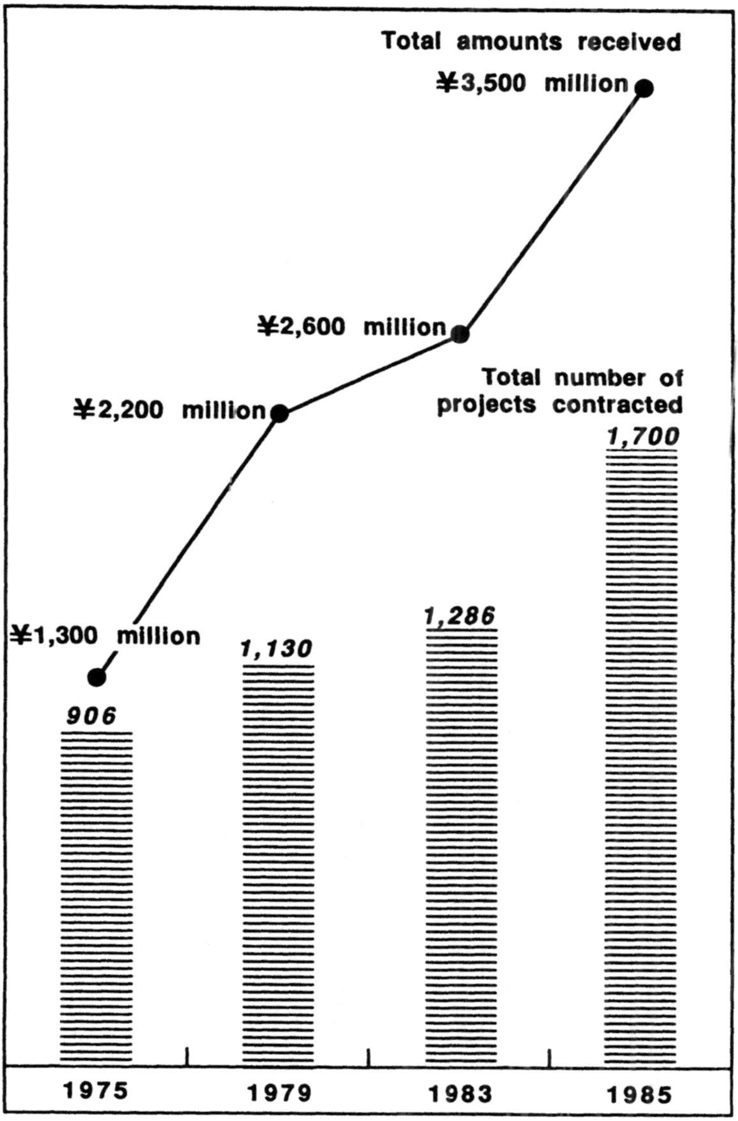

Table A3.1 Joint research with industry, 1986, Japan

			Funds (yen)	
Classification	*Research subject*	*Participants*	*NIT*	*Corporation*
A	Growth of multilayer thin-film crystal by MOCVD method and its application to photo-electric conversion device	Professor Masayoshi Umeno Assistant Shiro Sakai	6,346,000	9,800,000
A	Development of finishing construction method to improve durability of reinforced concrete structure	Professor Tatsuo Okajima Associate professor Shigeru Wakayama	1,600,000	3,000,000
A	Optimization of fine ceramics manufacturing process	Professor Suguru Suzuki Assistant Minoru Takahashi	1,111,000	1,000,000
A	Study on image reconstruction for 3-dimensional CT	Professor Nobuo Suzumura Associate professor Akira Iwata	1,056,000	1,300,000
A	Study of graphic processing software	Professor Naohiro Ishii Assistant [illegible]	891,000	1,000,000

Table A3.1 — continued

A	Development of exposed fixed pier and column-beam junction for steel structure	Professor Yasunaga Fukuchi	772,000	1,500,000
B	Performance improvement of pulse combustion burner and development of low-noise valve-less pulse combustion burner	Professor Shigeki Yamaguchi Associate professor Norio Oiwa	149,000	5,000,0000
C	Development of robot simulation system by next-generation 3-dimensional CAD	Associate professor Hideo Fujimoto	6,000	—
C	Study of permeability of polypeptide membrane	Professor Akira Takizawa Assistant Takatoshi Kinoshita	33,000	—

Source: Compiled by the author.

and the manufacturer. For example, the government, as the user, and industry have a long history of co-operation in the field of electrical communication. This co-operation has been very strong and has made a great contribution to the development of today's advanced technology. Many joint research projects have been organized in the fields of microwave communications, electronic switching systems, VLSI, and so on. Many corporations often participate in joint research. Thus the key to successful joint research is the co-ordination and management of the research programme, as well as strong leadership and a spirit of co-operation. The great progress in the electrical communications field in Japan was accomplished by determined men with strong leadership qualities both in research and management, and the manufacturers' willingness and efforts on behalf of co-operation and standardization.

Apart from joint research, researchers in universities may receive contract research funds that are offered by outside sources such as industry to have scientists carry out research on behalf of the sponsors (see Figure A3.5).

As a similar means, the contract research system allows for engineers and researchers from private corporations and public institutions to go to the universities to improve their knowledge through participation in university research. Donations are offered to the universities by private corporations. While the sponsor assigns the funds to particular scientists and their research subjects, no direct outcome from the research is expected, unlike joint research and contract research (see Figure A3.5).

The ideal way of university–industry co-operation

Private corporations often expect concrete results of research from universities in a short period of time. However, when considering the university's future development of technology, unlike the viewpoint of a corporation, it is important to take up and promote more long-term basic research subjects. In general, scientists in universities are engaged in research activities stimulated by their own interests; but it is not sufficient nor is it carried out systematically. In Japan, where industrial R&D has a strong base, such basic research must be promoted by the close co-operation of industry and universities.

Introduced here as an example of such university–industry co-operation is the newly established Advanced Telecommunications Research Institute International. This institution was established by the joint investment of industry to promote basic and advanced

Figure A3.6 Advanced Telecommunications Research Institute International and R&D Company

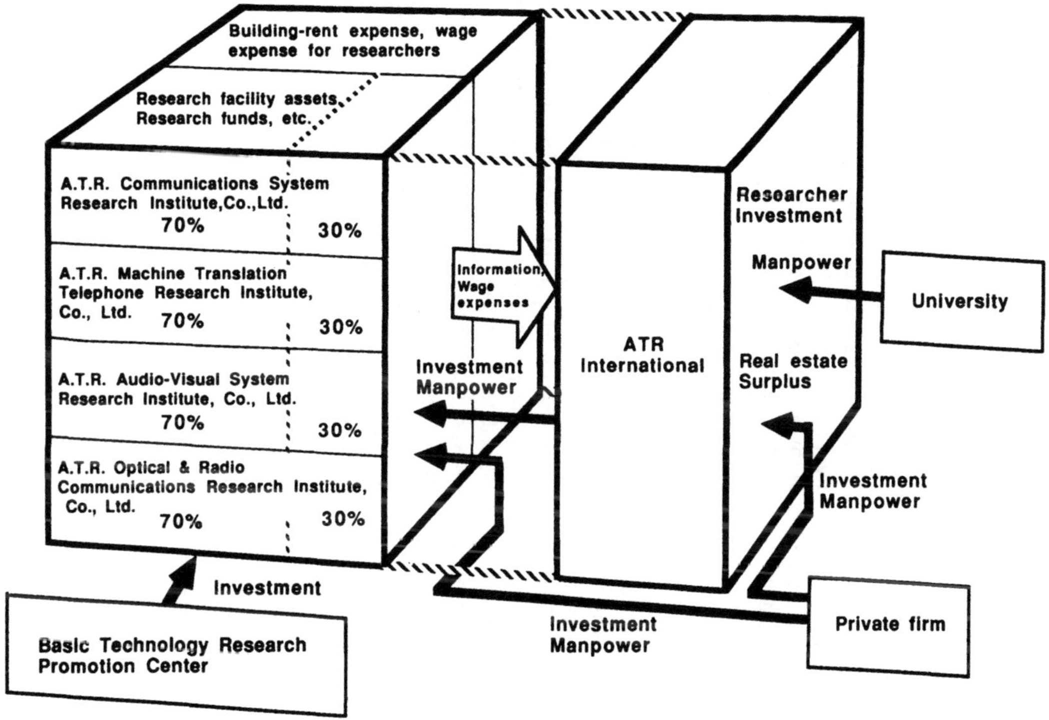

technology in the field of electrical communications, by installing state-of-the-art research facilities and attracting researchers from wide circles of industry and universities. No such large-scale basic institution had ever been established before, and fruitful results are expected (see Figure A3.6).

The system of joint research has been making sound progress and is expected to be improved and expanded in the future. This is because industry has a growing need for joint research. But industry is not in a position to have a full understanding of the research activities conducted in universities, and some measures to ensure better communications between university and industry are needed. Better communications between both parties will also be beneficial to the universities, in that the research activities will be vitalized. Recently vigorous efforts have been made to organize co-operation of local industry and universities in many districts of Japan.

In the district around Nagoya, representatives from academic, industrial, and administrative bodies have had discussions on this issue. The amount of industrial output from this district is unparallelled in Japan, but advanced technology is not significant in this area. And there is a pressing need to introduce high technology into the existing industries in order to improve the industrial structure of this district for its future development. For this purpose, plans have been proposed and are presently under study such as training of engineers in advanced technologies, establishment of a liaison and co-ordination body between the university and industry, establishment of a scientific research promotion fund, concentration of higher education and industry in an area, and so on. It is regarded as most beneficial and effective to promoting university–industry co-operation.

Selected bibliography

ACARD (Advisory Council for Applied Research & Development) (1983) *Improving the Research links between Higher Education and Industry*. London: HMSO.

Bolton, J. E. (1971) *Small Firms: Report of the Commission of Inquiry on Small Firms* (Cmnd 4811). London: HMSO.

Bragg, S. L. (1982) 'Technology transfer', *Measurement and Control* 16 (January 1983): 27–31.

Currie, Jean (1985) *Science Parks in Britain: Their Role in the late 1980s*. Cardiff: Currie Sunman Publications.

(1986) *New Technology — the Wealth Creator? University/Industrial Research — Benefits and Prospects. The Case of UWIST*. Cardiff: CSP Economic Publications.

Dalton, Ian (1986) 'Progress of science parks'. Paper presented to UKSPA Annual Conference 1986, 10 December 1986, London.

Debenham, Tewson, and Chinnocks (1983) *High Tech: Myths and Realities*. London: Debenham, Tewson and Chinnocks.

DES (1968) 'High technology training needs in North Wales'. Report compiled by NEWTECH (Clwyd) Ltd. for the Department of Education and Science under the Pick-up Programme, Newtech Clwyd Ltd, North Wales.

DTI (1985) *Science and Technology Report, 1983–84*. London: Department of Trade and Industry.

Economist Intelligence Unit (EIU) (1985) *Universities and Industry*. Special Report no. 213. London.

EEC (1986) 'Plan for the transnational development of the supporting infrastructure for innovation and technology transfer'. *EEC Newsletter* (ISSN 0255–0806) July 1985.

Ettinger, J. S. (1985) *Overview of Belgian Entrepreneurial Environment*. An international seminar held in London on 2–3 October 1985. Luxembourg: TII.

Goddard, J. B. and Thwaites, A. T. (1983) 'Science parks in the context of national, regional and local technology policy'. *The Planner* 69, no. 1.

Hepworth, G. M. (1985) 'The scientific and technical information scene in Japan: recent trends and developments'. Paper presented at a seminar on Japanese scientific, technical and commercial information. London: British Library, Reference Division.

Kikuchi, M. (1986) 'Co-operation of industry in Japan with higher education'. Paper presented at a seminar on university–industry interaction. Shinagawa-Ku, Tokyo.

Lieber, H. W. (1985) *German Universities and Technical Colleges as Partners of SMEs*. An international seminar held in London on 2–3 October 1985. Luxembourg: TII.

Loveland, R. (1986) 'Co-operation of industry in Europe with higher education'. Paper presented at a seminar on university–industry interaction. Shinagawa-Ku, Tokyo.

Mackensie, I. and Jones, R. R. (1986) 'Universities and industry — new opportunities from collaboration with UK universities and polytechnics'. Special Report no 213. London: Economic Intelligence Unit.

Merrison, A. (1982) *Report of a Joint Working Party on the Support of University Scientific Research* (Cmnd 8567). London: HMSO.

Monck, C. S. P. (1986) *Science Park Directory*. UK Science Park Association.

Monck, Porter, Quintas, Storey, and Wynarczyk (1986) *Science Parks and the Growth of High-Technology Firms*. London: Croom Helm.

Moore, B. and Spires, R. (1983) *The Experience of the Cambridge Science Park*. OECD Workshop, 24–27 October 1983. Paris.

Morita, M. (1986) 'Co-operation of industry in Japan with higher education'. Paper presented at seminar on university–industry interaction. Shinagawa-Ku, Tokyo.

National Economic Development Office (1984) *Competence and Competition — Training and Education in the Federal Republic of Germany, the United States and Japan*. London.

National Science Foundation (1982) 'University–industry research relationships, myths, realities and potentials'. Fourteenth Annual Report of the NSB. Washington, D.C.

Oakey, R. (1984) *High Technology Small Firms: Innovation and Regional Development in Britain and the United States*. London: Frances Pinter.

OECD (1982) *Innovation in Small and Medium Firms*. Paris.

(1986) *Science and Technology Indicators. no. 2: R&D Innovation and Competitiveness*. Paris.

Ohta, M. (1986) 'Co-operation of higher education in Japan with industry'. Paper presented at a seminar on university–industry interaction. Shinagawa-Ku, Tokyo.

Parkes, L. R. (1983) 'Paved with the best of intentions: the ACARD/ABRC report debated'. *The Metallurgist and Materials Technologist Journal* (1983): 618–21.

P. A. Technology(1985) *Attitudes to New Technology — an International Survey*. Herts: MORI.

Pedinelli, A. (1985) *The Opening of Universities to Small and Medium-sized Enterprises*. International seminar held in London on 2–3 October 1985 at the Institution of Mechanical Engineers. Luxembourg: TII.

Phillips, G. O. (1982) *Era of Change*. British Broadcasting Corporation Lecture (ISBN 563 201630).
Quinn, J. J. (1985) 'Information and the industrial chemist'. *Chemistry in Britain* (August 1985): 738–39.
Sassoon, J. (1985) 'Japanese scientific and technical publications'. Paper presented at a seminar on Japanese scientific, technical and commercial information. London: British Library, Reference Division.
Segal, Quine, and Wicksteed (1985) *The Cambridge Phenomenon: The Growth of High-Technology Industry in a University Town*. Cambridge: Segal, Quine, Wicksteed.
Select Committee on Science and Technology (1975) *First Report from the Select Committee on Science and Technology*. London: HMSO.
Shimizu, J. (1986) 'Co-operation of higher education in Japan with industry'. Paper presented at a seminar on university–industry interaction. Shinagawa-Ku, Tokyo.
Society of Education Officers. *Key Issues for industry and education*, No. 3. London, 1986.
Suzuki, S. (1986) 'An international viewpoint'. Paper presented at a seminar on university–industry interaction. Shinagawa-Ku, Tokyo.
Uchida, M. (1986) 'Economic development and industrial property — Japan's experience. *AIPPI Journal*: 59–76.
UKSPA (1986) 'Summary of science parks in the UK'. Paper presented at UK Science Park Association, Annual Conference 1986, 10 December 1986. London.
Wield, D. and Quintas, P. (1986) 'The technological relevance of science parks'. Paper presented at UKSPA Annual Conference 1986, 10 December 1986. London.
Wijnde, W. J. (1985) *Cooperation between the Fachhochschule of Aachen (W. Germany), The WTS of Heerlen (Netherlands), and DSM/ Stamicarbon (Netherlands)*. International seminar held in London on 2–3 October 1985. Luxembourg: TII.

Index

Note: Page references printed in italics refer to pages where tables appear.

For Product Safety Concerns and Information please contact our EU representative GPSR@taylorandfrancis.com Taylor & Francis Verlag GmbH, Kaufingerstraße 24, 80331 München, Germany

Batch number: 10725802

Printed by Printforce, the Netherlands